Undergraduate Topics in Computer Science

'Undergraduate Topics in Computer Science' (UTiCS) delivers high-quality instructional content for undergraduates studying in all areas of computing and information science. From core foundational and theoretical material to final-year topics and applications, UTiCS books take a fresh, concise, and modern approach and are ideal for self-study or for a one- or two-semester course. The texts are all authored by established experts in their fields, reviewed by an international advisory board, and contain numerous examples and problems, many of which include fully worked solutions.

The UTiCS concept relies on high-quality, concise books in softback format, and generally a maximum of 275–300 pages. For undergraduate textbooks that are likely to be longer, more expository, Springer continues to offer the highly regarded Texts in Computer Science series, to which we refer potential authors.

More information about this series at http://www.springer.com/series/7592

Alexei Sourin

Making Images
with Mathematics

Alexei Sourin
School of Computer Science
and Engineering
Nanyang Technological University
Singapore, Singapore

ISSN 1863-7310   ISSN 2197-1781   (electronic)
Undergraduate Topics in Computer Science
ISBN 978-3-030-69834-8   ISBN 978-3-030-69835-5   (eBook)
https://doi.org/10.1007/978-3-030-69835-5

This Springer imprint is published by the registered company Springer Nature Switzerland AG
The registered company address is: Gewerbestrasse 11, 6330 Cham, Switzerland

To my wife Olga

Preface

Aim of the Book

Information visualization creates images from abstract data by interpreting them geometrically. It may require skills of making images with raw mathematics while the graphics content is now mostly created by using sophisticated licensed software. As a result, fewer and fewer developers are capable of using procedurally based visualization in which mathematical formulas are used for defining complex geometric shapes, transformations, and motions as well as for coloring the geometry.

The book explains how to see geometry and colors beyond simple mathematical formulas and teaches how to represent geometric shapes and motions from first principles by digital sampling mathematical functions.

The book may serve as a self-contained text for a one-semester computer graphics and visualization course for computer science and engineering students, as well as a reference manual for researchers and developers.

Book Organization

The book has seven chapters.

The Chap. 1 "From Ancient Greeks to Pixels" explains how we see the world and how the computer makes images. Beginning with Ancient Greek Geometry, it travels to modern geometry, introduces the subject of computer graphics and visualization, explains how the graphics pipeline works, and how a geometric point turns into a color spot on the computer screen.

The Chap. 2 "Defining Geometric Shapes" presents the mathematical foundations of shape modeling. Curves, surfaces, and solid objects are considered as a set of points which are obtained by sampling various types of mathematical functions. Using the concept of sweeping, many varieties of shapes are defined based on only a few simple foundation principles.

The Chap. 3 "Transformations" considers how the same formulas, used for making shapes, can define their transformations. The rationale for using matrix transformations is explained, and affine and projection matrix transformations are presented. Generalization of geometric sweeping implemented with matrices is further discussed.

The Chap. 4 "Motions" explains how the previously introduced mathematical formulas, defining shapes and transformation matrices, can be extended to time-dependent models of moving shapes. Motions of rigid shapes and shape morphing transformations are considered. Besides pseudo-physical motions, definitions based on Newtonian physics are also introduced.

In the Chap. 5 "Adding Visual Appearance to Geometry", we consider how colors can be added to geometry, and how its photorealistic appearance can be achieved. The formulas, previously used for defining geometry, now will define variable colors as a new modality of immersion into the world of geometric definitions.

In the Chap. 6 "Putting Everything Together", the ways of making interactive, real-time, and immersive visualization environments are considered including technical and physiological design and implementation issues. Still the same transformations, and actually the same basic mathematical principles, will be used in the fast visualization methods.

Finally, the Chap. 7 "Let's Draw" introduces to the reader a few commonly used freeware software tools—OpenGL, POV-Ray, VRML, and X3D—which will let the readers apply theoretical principles into practice without requesting expensive hardware and software solutions. Also, the readers will learn how immersive visual mathematics can be implemented using the function-based extension of VRML and X3D, which allows for defining geometric shapes and their appearances with analytical functions. Finally, the Shape Explorer tool will be presented to the reader as a multi-platform companion viewer for all the examples used in the book.

Acknowledgments

I am very grateful to the School of Computer Science and Engineering of Nanyang Technological University for the pleasure of doing research and teaching, which eventually resulted in writing this book. I am very thankful to my students and colleagues for their comments and suggestions.

Last but not least, I would like to thank Helen Desmond, Editor for Computer Science of Springer Nature, who offered continuous encouragement and support.

Singapore, Singapore   Alexei Sourin
2021

Contents

From Ancient Greeks to Pixels

1.1 Drawing with Computer

1.1.1 How We See the World

We do not see with eyes. We visually perceive the world *with our brain through our eyes* which have photoreceptors layers (or light-sensitive cells) at the back that is called the retina [1]. These photorcccptors produce electrical signals when they are illuminated by light, or to be scientifically precise, excited by what we call red, green, and blue parts of the visible light spectrum. Photoreceptors called *cones* are responsible for color vision, *rods*—for monochrome night vision, and *ganglion cells* for the peripheral vision which mostly detect motion rather than colors. All the electrical signals produced by the photoreceptors are carried out by the optical nerve into the back part of our brain—the occipital lobe—which interprets them as images. The brain merges two different visual signals obtained from the eyes into what we understand as a three-dimensional (3D) image with depth. Only living creatures which have eyes positioned on the front of their heads and looking forward have this binocular stereo vision. These are usually predators—tigers, eagles, and …humans—that need the sense of depth to estimate a distance to the prey which they are hunting for food, while the preys, in turn, need a vision with as wide angle as possible, and, therefore, they have eye positioned on opposite sides of their heads (e.g. cows, fish). It is interesting that there is a growing loss of binocular vision in urban societies (*stereo blindness*) since we do not have to hunt for food anymore but merely need to know how to read a price tag in food stores. Thus, many people have a vision disorder with one leading eye while the other 'slave' eye only provides for a wider view angle (*amblyopia* or *lazy eye*). There are many online resources to instantly check how good your depth perception is [2].

The light, which comes to our eyes, is a combination of the light coming directly from the light source (when we look straight at it), the light reflected from the surfaces of objects, and the refracted light passed through translucent (semi-transparent) objects. The reaction time of the rods in the retina is around

© The Author(s), under exclusive license to Springer Nature Switzerland AG 2021
A. Sourin, *Making Images with Mathematics*, Undergraduate Topics
in Computer Science, https://doi.org/10.1007/978-3-030-69835-5_1

0.25 s while cones react about four times faster (~ 0.06 s). The combined speed of reaction creates the phenomenon called *persistence of vision*, which allows us to continue seeing the object for some time after the rays of light are no longer coming from it. Our brain then integrates discrete images of moving object into what appears to us as a continuously moving object [3].

1.1.2   Displaying Images

Visualization requires the replacement of the real 3D light reflection/refraction with merely displaying the resulting images to our eyes on some media. These can be images projected onto screens or displayed on various video monitors.

The first photo-realistic visualization was achieved with the invention of photography which was used together with the so-called magic lantern (also known as slide projectors) and later with cinema projectors. The believable immersion was achieved by the large size of the screen occupying peripheral vision of the eyes. Also, the images were mostly displayed in dark rooms that improved concentration on them by psychological elimination of the surrounding world.

Film projectors (Fig. 1.1) may display as little as 16 images per second to achieve the effect of persistence of vision, which is in fact the cones reaction time of ~ 0.06 s/image. While this minimal projection speed was experimentally determined, 24 frames/s had eventually become used as a projection speed in a pursuit to achieve a compromise between the quality of projection and the high cost of films. The film projection is achieved by stepwise advancing of the film synchronously with rotation of the propeller-like shutter. Each frame is illuminated by a bright lamp when it is positioned in front of the projection lens, and the light is blocked when the film is being advanced to the next frame. A possibility of the reduction of the speed of projection to as low as 16 frames/s is also supported by the effect of *motion blur*, which can be observed on each individual frame of a film. If a movie is paused during a high-action scene with lots of movement, the moving objects will mostly look blurred in the direction of their movement, but when it is played again, the movie will look good and sharp. The motion blur is created due to the way how the movie film is made. While shooting the movie, each individual frame is exposed for a certain time set by the shutter speed of the camera during which the moving objects continue to move thus creating multiple copies of themselves on the frame. This blur then acts like a speed gradient to prompt the brain about the direction and speed of motion of the objects. In computer animation, therefore, more frames per second have to be displayed to replace the absence of motion blur since each image is created with a computer rather than captured with a camera. The search for the perfect projection speed still continues. Thus, in recent 3D movies displayed in the early 2000s much higher projection speed of 144 frames/sec was used. It is obtained by doubling 24 frames/sec, since images for the left and right eye have to be shown subsequently, and furthermore each frame is repeated 3 times for better stability, i.e. $24 \times 2 \times 3 = 144$.

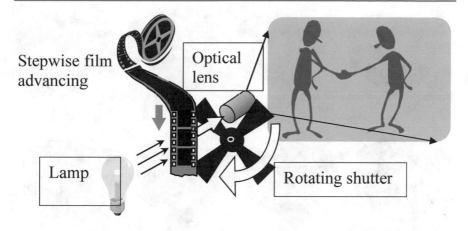

Fig. 1.1 Film projector

1.1.3 Computer Monitors

The invention of television introduced *Cathode-Ray Tubes* (CRTs) which were later used with computers as the first video monitors. The main part of the CRT is a large glass bulb. At one side of it, a so-called *electron gun* is placed, while the other wider side of it is covered inside with a thin *phosphor layer*. The idea of CRT is that if the electrons are directed toward the phosphor-coated screen and strike it at a high speed, the phosphor will emit a short flash of light at the excited point. This flash of light is seen from outside the bulb as a light spot on the screen. Photon excitation decays fast—within 10–30 ms. While the electrons are constantly directed to the same spot on the screen, the light spot remains visible. Drawing images on the screen becomes possible if the electron beam is frequently directed at different spots of it constantly revisiting, or scanning, them to prevent the emitted light from decaying. Since each point will flash for only a short time, the image has to be redrawn again and again, typically at least 60 times per second, to make it look steady on the screen. When another pattern is to be drawn alongside the first one, the electron beam has to be cut off momentarily and then restarted from the new initial point. This motion of the electron beam is provided by various deflection and focusing systems. Depending on how the image is scanned, the CRT can be used in random- or raster-scan monitors (Fig. 1.2).

Random-scan monitors display images in a way how we draw with a pen: curve by curve. The advantage of this design is that high-resolution contour-based images can be displayed on relatively large screens, which is a requirement of computer-assisted design and architectural applications—prominent uses of computer graphics. Nowadays, random-scan images can be seen in various laser-fountain public displays where in place of electron beam, colored laser beams are used to draw images on water mists. The serious drawback of the random-scan drawing is that the necessity to redraw or refresh an image many times per second

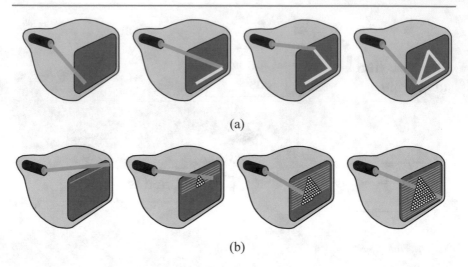

(a)

(b)

Fig. 1.2 a Random-scan and **b** raster-scan monitors

imposes restrictions on its size in terms of the total length of a curve which is to be drawn. As such, after a certain size of the image, it will begin to *flicker* since there will not be enough time to refresh the whole image fast enough to avoid fading of the points lit up first. Also, the random-scan design does not easily allow for displaying shaded images.

The disadvantages of the random-scan monitors have been solved in *raster-scan monitors* which form images by individual colored dots called *pixels* (short for *pic*ture *el*ements). These pixels form a raster matrix on the screen with typical resolutions of 800 × 600, 1024 × 768, 1280 × 960, etc. Each pixel is formed by 3 *sub-pixels* made of different types of phosphor: one producing red color, another one—green, and the third one—blue. To display an image on the screen, the electron beam has to scan every line of the raster matrix in sequence from top to bottom to light up those pixels which have to be in the foreground, while leaving unchanged those pixels which form the background color. Note that in this case, it does not matter how complex the image is—the electron beam has to scan through the whole screen anyway. On one hand, this method reduces the quality of the image, since it consists of a limited number of pixels, but on the other hand it has an advantage that it does not matter how many geometric points, curves, or polygons have contributed to the image on the screen—the refresh time remains the same provided all these geometric images have been converted to the special raster memory where the values of the pixels are stored. The problem of *rasterization* or conversion of geometric shapes into their raster representation is central to basic computer graphics. The larger the size of a screen is, the longer time will be required to refresh the raster image on it. It may create an undesirable effect of flickering when some pixels start to fade before they are refreshed.

Liquid Crystal Displays (LCDs), which replaced CRTs, use a different technology of turning the pixels on and off [4]. *Liquid crystals* twist or change their orientation to varying degrees under an electric current depending on the voltage applied. LCDs use these liquid crystals because they react predictably to electric current in such a way as to control polarized light passage. The light is polarized with a special polarized glass. An LCD, which can show colors, has three sub-pixels with red, green, and blue color filters to create each color pixel. The advantage of LCDs is that the pixels remain lit while the voltage is being applied to the respective crystals. As such, there is no flickering on LCD panels however to maintain interactivity the images still have to be refreshed at least 40 times/s.

1.1.4   Storing Images in Computers

There is a special memory where the information about pixel colors is stored. It is called a *frame buffer*. For each pixel, a certain number of bits of color information is allocated. The number of colors which can be displayed with a given frame buffer, equals to 2^n, where n is the number of bits per color used in this frame buffer. Thus, 8 bits per color allow for defining 256 different colors, while 24 bits create 16,777,216 colors. On some monitors, we can change the number of colors versus the raster image resolution so that when we increase the number of colors, the resolution decreases, and vice versa. The method when colors are directly encoded into the bits allocated to each pixel is often called *True Color Mode*. In that case, each color is defined by three integer numbers stating the amount of red, green, and blue in the pixel (see Fig. 1.3).

It may happen that the total number of colors in the True Color Mode is still less than what is required for the particular application. For such cases, as well as for the convenience of remembering a color by its name or reference number rather than by

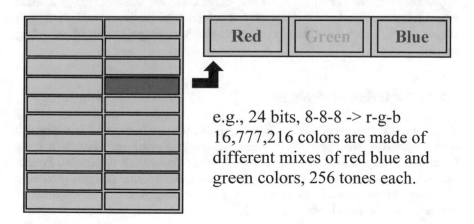

e.g., 24 bits, 8-8-8 -> r-g-b 16,777,216 colors are made of different mixes of red blue and green colors, 256 tones each.

Fig. 1.3 Frame buffer in a true color mode

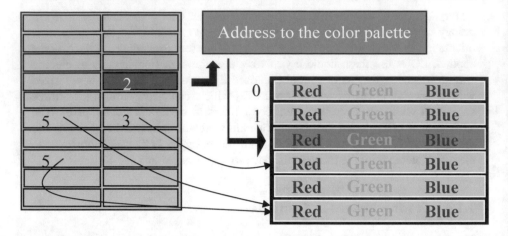

Fig. 1.4 Frame buffer with a Color Palette

its components, an *Indexed Color Mode* can be used as an alternative to the True Color Mode. In this case, the bits associated with each pixel are used for addressing other parts of the display memory, where the actual colors are encoded. In this memory, or *color palette*, the number of bits allocated to each color maybe even larger than the number of bits used for encoding the color address or color index (Fig. 1.4). The color palettes can be reloaded without changing the content of the frame buffer which will recolor the image.

1.2   From "Earth Measuring" to Computer Graphics

When drawing, we sketch points, lines, curves, planes, surfaces, etc. All these words are geometrical terms. Geometry is a branch of mathematics dealing with shapes, sizes, relative position of objects, and various properties of modeling spaces.

1.2.1   Evolution of Geometry

Geometry was invented in ancient Greece. The Greek word 'geometria' (γεωμετρία) comes from two words *geo-* "earth" and *-metron* "measurement". For the ancient Greeks geometry was more than a science about "measuring the earth". Perfect geometric proportions (or so-called Golden Ratio), architectural designs, geometric ornaments in decorations—these are just a few examples where applications of geometry can be found, and which still exist in our world. Geometry in ancient Greece was more than a field of mathematics but rather an attempt to explain the universe. Thales of Miletus (635-543BC) used geometry to calculate the heights of pyramids and distances of ships from the shore, and he also applied deductive reasoning to geometry. Pythagoras (582-496BC), besides studying mathematics, music, and philosophy, discovered most of what high school students learn today in their geometry courses. In the philosophy of Plato (427-347BC) each of four classical elements of the world—earth, air, water, and fire—were associated with four regular solids: cube, octahedron, icosahedron, and tetrahedron. It was inscribed at the door of his academy "Let none ignorant of geometry enter here!" Aristotle's (384-322BC) methods of reasoning were used in deductive proofs of geometry. Euclid (325-265BC) presented geometry in an ideal axiomatic form known now as Euclidean geometry. More about Greek geometry can be found in [5].

However, through all these centuries, geometric tasks were performed only by drawing and measurements. Algebra was not used in solving geometric problems. The methods used for solving practical geometric problems in the ancient Greek way are still very much used. For example, as illustrated in Fig. 1.5, a center of a circle can be found by drawing $line_1$ and $line_2$, followed by drawing two intersecting arcs with radius rad_1 and two intersecting arcs with radius rad_2, and finally by drawing two lines $line_3$ and $line_4$ which intersection point is the center of the circle.

Greek geometry reached its pinnacle during the Archimedes time (287-212BC) —the greatest of the Greek and world mathematicians. He proposed methods very similar to the coordinate systems of analytic geometry but still lacking algebraic notations.

Then, century by century, geometry and algebra existed side-by-side however not interfering and contributing to each other. Conversion from Roman numerals [6] to our current number system which we call Hindu-Arabic numerals (invented in India and then brought to Europe by Middle-Eastern mathematicians [7]) helped to move algebra in the directions which were not considered before: negative numbers and zero number were introduced. Using decimal separator allowed for more efficient writing of fractional numbers compared to how it was done using Roman numerals.

Finally, a "geometric revolution" happened in the seventeenth century. In 1637, French scientist Rene Descartes published "La Geometrie" in which he proposed definition of geometric curves by their algebraic equations. It took more than

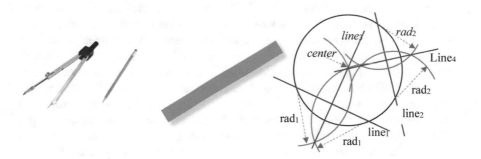

Fig. 1.5 Finding the center of a circle by drawing

10 years to be accepted by the scientific community, and only after Descartes translated his monograph into Latin, which was the language of international scientific communication at this time, under his Latinized name Cartesius (obtained from *Descartes* by removing prefix '*Des*' and attaching Latin suffix '*ius*'). Descartes invented system of measuring locations of points in space by signed numbers (called coordinates) which are used to uniquely determine the position of a point or other geometric elements in the geometric space. Since then we know the *Cartesian coordinate system* where coordinates are measured as distances from a point to the coordinate axes located at 90 degrees to each other. Together with Descartes, another scientist, Pierre de Fermat considered how starting with an algebraic equation to describe the geometric curve which satisfied it. This is how the bridge between the geometry and algebra was built. This was the time when other coordinate systems were also invented—polar, cylindrical, spherical, etc. This revolution in geometry eventually created many new sciences such as analytic geometry, linear algebra, complex analysis, differential geometry, calculus, and... eventually, *computer graphics and visualization, shape modelling, computer animation and virtual reality systems.*

1.2.2 Computer Graphics and Beyond

Computer Graphics [8] is a synthesis of images with a computer, while a converse process of analyzing images is called *Image Processing* [9]. Generally speaking, every work with a computer, which produces graphics images on the screen or a hardcopy of such images, can be called computer graphics. The hardcopy can be both a printed or plotted graphics image as well as a physical three-dimensional model made by a 3D printer. Computer graphics history dates back to the early 1950s when the computers were first used to control CRT monitors. During the next decade, from 1960, the main basic computer graphics algorithms were devised, and special display processors were developed. The 1970s was known as the raster graphics revolution decade. Relatively cheap raster displays brought computer graphics to practically every office and to many homes. With raster displays, computer graphics left sophisticated research labs and spread all over the world. During the next decade, from 1980 to 1990, special computer graphics computers or graphics workstations were developed. These workstations, designed specifically for solving advanced computer graphics problems, enabled computer graphics to mature in the areas of computer science and engineering, and initiated the development of uses and areas of application briefly mentioned above. The current stage of computer graphics features a rapid development of multimedia, virtual and augmented reality systems, volume graphics and shaders (little graphics programs) developed on graphics processing units (GPU), which bring visual realism as well as immersion to a principally new level. The ability to create virtual worlds inside a computer and to even cheat our senses, so that the feeling of "being there" becomes very real, is the current stage of computer graphics, and even off-the-shelf personal computers can be used for such applications.

Visualization [10] considers how to form an image. It can be displaying of common geometry used in everyday life (e.g. plane, sphere, cube) using various graphics attributes or some data which may not even have an apparent visual appearance to be used (e.g. radiation, magnetism, soundwaves). Information visualization involves various mathematical models, and it has become important in modern science and technology especially in cases where simple plots and diagrams are insufficient. It has applications across many disciplines including visual analytics where analytic reasoning is supported by various methods of interactive visualization.

The narrow meaning of the term computer graphics may often refer to only *basic computer graphics*, while we also have subjects of *interactive computer graphics* and *graphical-user interfaces, geometric and shape modeling, computer animation, real-time rendering* and *virtual reality systems*.

Basic computer graphics considers drawing algorithms and procedures. Most often, these are procedures which are able to draw such objects as lines, circles, shaded polygons, and text strings.

Interactive computer graphics [11] allows us to communicate with the computer through the images on the screen. These images are parts of the graphical-user interfaces, such as icons, menus, 2D and 3D cursors, etc. They can also be located on the screen with interactive tools, such as computer mouse, trackball, joystick, touchpad, etc., and then be moved, transformed, regrouped, or modified in one way or another. In either case, interactive computer graphics assumes that behind color spots constituting an image on the screen, there is also a data model accessible through different parts of the image so that even a three-dimensional object displayed on the screen can be manipulated interactively.

Geometric and shape modeling [12] is about defining a mathematical model of a two- or three-dimensional object including its geometry and sometimes even physical properties of an object. Geometric models may have different graphical representations. For example, 3D models of moving particles can be defined with just coordinates of the moving points, while their graphics representations may look like either little spheres, or small triangles, or just colored dots.

Computer animation [13, 14] considers moving objects and the ways of making geometric models time dependent. Computer animation spans from computer-assisted animation, which allows cartoonists to save a lot of time on producing so-called "in-betweens", to intelligent animation, which deals with virtual actors that have behavior and are capable of moving on their own in virtual environments.

Real-time rendering [15] expands computer animation into our time domain, when the computer becomes capable of making interactive changes to images with anticipation of our control of the motion in the computer-generated world. This part of computer graphics deals with different simulators, such as driving and flight simulators and three-dimensional computer games.

Virtual reality systems [16, 17] are built on the most recent achievements in interactive computer graphics, shape modeling, computer animation, and real-time rendering. These systems offer us an alternative world inside a computer, which we

can communicate with a great degree of immersion and often with a haptic feed-
back. The objects of virtual reality act like real objects rather their graphics images.

Immersion [14] is a feeling of being physically present in some environment
simulated in the computer. Immersion is required in virtual reality systems where it
is achieved by smart combination of special visualization methods and interactive
hardware devices. Immersion however can be used to facilitate solving of information visualization as well by providing multimodal interaction with the data
mimicking the way how we interact in real world. Thus, working on some visual
analytic problem, we may use different ways of visualizing data on high resolution
displays while interacting with the analytic software using a common mouse
together with 3D hand-tracking and eye-tracking cameras as well as
voice-communication software.

1.3 We Need Digits to Draw with Computer

Computers use digits. Therefore, to visualize the real world, which is continuous
and infinite, we have to digitize it, and this will be a discrete representation which
will sample objects to be displayed within the computer precision limits. The digits
can be obtained using coordinate systems as a frame of reference and based on
some mathematical models (formulas, equations, procedures, algorithms) which
then have to be digitally sampled with a various precisions limited by the precision
of the computer. Therefore, it is always a task of interpolation or approximation
based on a digital model used in the computer.

Thus, we may represent a circle by its algebraic equation that is then have to be
sampled to produce coordinates of many points on the circle which will be connected with straight lines to interpolate the shape of the circle (Fig. 1.6a, b). The
same equation of a circle can be, however, sampled with a fewer points to eventually display very different shapes (Fig. 1.6c, d).

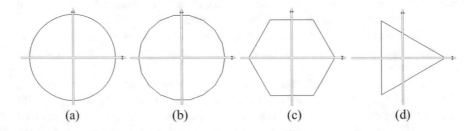

(a)   (b)   (c)   (d)

Fig. 1.6 Curve interpolation based on an algebraic equation of a circle with (**a**) 72 sampling point
will nicely interpolate a circle, (**b**) 18 points will produce rather coarse interpolation which still can
be recognized as a circle, (**c**) 6 sampling points will make a hexagon, and (**d**) 3 points will display
an equilateral triangle

1.3.1 2D Cartesian Coordinates

The object to be drawn has to be defined in one of the coordinate systems. Usually, these are two- or three-dimensional Cartesian, Skew, Polar, Cylindrical or Spherical Coordinate Systems.

The number of coordinates matches the dimension of space. If this is a 2-dimensional plane (2D), then two coordinates have to be used. In a 3-dimensional space (3D), three coordinates have to define location of every point, etc. Einstein's general theory of relativity considers time as yet another dimension together with 3 spatial dimensions, hence it will require 4 coordinates. Other modern physics theories explain the universe with 10, 11 and even 26 dimensions. In this book, however, we will use up to 3 spatial dimensions while the time will be used as yet another but different dimension.

Cartesian coordinate system (Fig. 1.7a) is defined by selecting a point—the origin—on a plane through which two orthogonal axes (directed lines) have to be drawn. The direction of the axes shows on which side of the origin the positive coordinates are located. The coordinates are then measured as signed distances from a point to be located to the coordinate axes. These distances are in fact measured along the lines cast from the point toward the coordinate axes while being parallel to them. We then have to define the order in which the coordinates have to be listed to uniquely define each point on the plane. That means we have to decide which coordinate axis has to be considered first and which second. The so-called "right-hand rule" is used for assigning this order to the coordinate axes. While the right hand is placed in front of the coordinate plane with a thumb pointing at you, the fingers will curl from the first to the second axis within the 90 degrees angle between them. If we call the first axis X and the second axis Y, the XY coordinate system thus defined will be called right-handed Cartesian coordinate system (Fig. 1.7a). Opposite direction of rotation will define what is called left-handed coordinate system, i.e. YX in our case. The coordinates (distances) can be positive and negative real numbers, and in each right- and left-handed Cartesian coordinate systems these two numbers uniquely define location of each point on a plane. For example, the same point in Fig. 1.7a will be defined by coordinates (3, 5) in the right-handed XY coordinate system and (5, 3) in the left-handed YZ coordinate system. It is not the only way to use the coordinate axes located at 90 degrees to each other as in the Cartesian coordinate systems. This location is convenient in many practical applications but in some application problems other angles between the axes can be used. Still the coordinates can be defined as linear signed distances to the coordinate axes measured along the lines parallel to these axes. These coordinate systems are called *skew systems* (Fig. 1.7b).

1.3.2 Polar Coordinates

In polar coordinates (Fig. 1.7c), each point on a plane is defined by a distance r to it from a certain point that is called the origin, as well as an angle θ measured from an

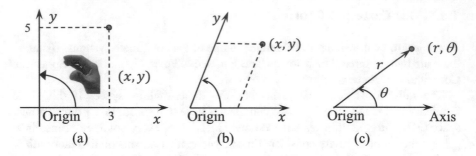

Fig. 1.7 Two-dimensional coordinate systems: **a** 2D Cartesian right-handed, **b** 2D Skew (non-Cartesian, right-handed), and **c** Polar

axis passing through the origin and a ray cast through the origin to the point. However, while in Cartesian and skew coordinate systems the coordinates can be any real numbers yet uniquely defining locations of points, in polar coordinates we may have ambiguous definitions of the same point with infinite combinations of the distance and angle. Indeed, the same point can be defined with a certain distance r but with many angles θ, $\theta + 2\pi$, $\theta + 4\pi$, ... , $\theta + 2n\pi$, where n is any integer number (positive or negative). Moreover, allowing the distance to be a negative number, we may define the same point with coordinates $(r, \theta + 2\pi)$ or $(-r, \theta)$. This ambiguity of definitions can be minimized if we set limits to the values of the polar coordinates, e.g. only positive values of r and the angles within the domain $[0, 2\pi)$. However, as we will see in the next chapter, the ambiguity of the polar coordinates will allow us to define very sophisticated shapes which can be very difficult and not intuitive to define in other coordinates.

Conversion between Cartesian and polar coordinate system is based on using Pythagoras theorem and its trigonometric definitions (Fig. 1.8) where the shot sides of the right triangle are the Cartesian coordinates while the longer side is the polar distance coordinate. The angle is the polar angle θ.

Let's consider a few examples of conversions.

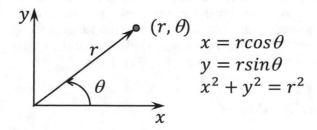

$$x = r\cos\theta$$
$$y = r\sin\theta$$
$$x^2 + y^2 = r^2$$

Fig. 1.8 Conversion between polar and Cartesian coordinates

To convert a straight line defined in Cartesian coordinate system XY by equation $y = 2x + 3$ to polar coordinates, we will follow the equations from Fig. 1.8 which will hold for every point on the straight line:

$$x = r \cos \alpha \qquad y = r \sin \alpha \tag{1.1}$$

Therefore:

$$r \sin \alpha = 2r \cos \alpha + 3$$
$$r = 3/(\sin \alpha - 2 \cos \alpha) \tag{1.2}$$

The domain for the polar angle is defined by the line slope angle (Fig. 1.9) which tangent is a coefficient in front of x:

$$\arctan(2) = 63.4 < \alpha < \arctan(2) + \pi = 1.107 + 3.141 \, \text{rad} = 243.4° \tag{1.3}$$

The origin centred circle with radius R will be defined in polar coordinates by $r = R$ with $\alpha \in [0, 2\pi)$, i.e. for any polar angle we display a point located at distance r from the origin. Note that the domain $[0, 2\pi]$ will make the equation such that we visit the same point twice when $\alpha = 0$ and 2π.

Fig. 1.9 The straight line in the Cartesian and polar coordinates

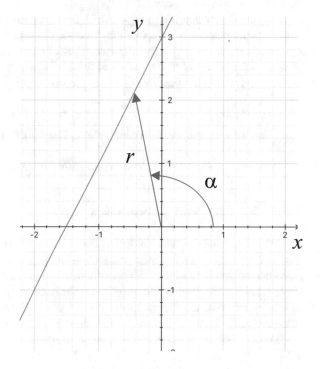

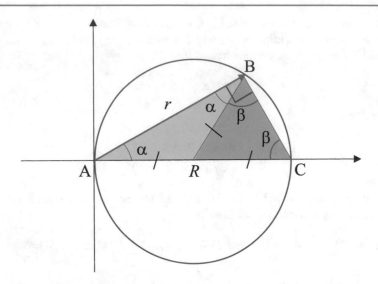

Fig. 1.10 A circle with the center at $(R, 0)$

The polar definition of a circle will become very different if its center is located at the Cartesian coordinates $(R, 0)$ (Fig. 1.10).

You may know that the triangle ABC, inscribed into the circle so that its longer side is its diameter, is a right triangle. One simple proof to it is to split this triangle into two isosceles triangles (having two sides of equal length) ABR and BCR. Two of their sides are equal to the radius of the circle, and the respective angles α and β are equal. Then, since the sum of the angles of any triangle is $180°$, we can write

$$\alpha + (\alpha + \beta) + \alpha = 180° \text{ which yields } 2(\alpha + \beta) = 180° \text{ or } \alpha + \beta = 90°$$
$$\alpha + (\alpha + \beta) + \beta = 180° \text{ which yields } 2(\alpha + \beta) = 180° \text{ or } \alpha + \beta = 90°.$$

Then, from the right triangle we can write $r = 2R\cos\alpha$ where $\alpha \in [-0.5\pi, 0.5\pi]$

1.3.3   3D Cartesian Coordinates

In three-dimensional space, Cartesian coordinates are created by adding one more axis orthogonal to the 2-dimensional coordinate plane. It also passes through the same origin, and the order of the coordinates is defined by the same right-hand rule (Fig. 1.11a): while curling fingers of the right hand show the direction from the first to the second axis, the thumb shows the direction of the third axis. These axes are often called X, Y and Z but, of course, any other names can be used for them.

Skew coordinate systems can be created in the same way by adding the third axis which passes through the same origin. Such coordinate systems are used, for example, in crystallography where there are natural inclination angles following the shape of crystals.

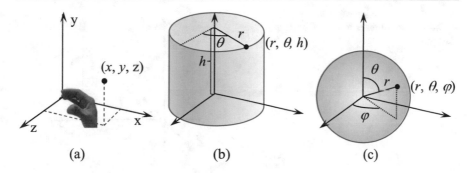

Fig. 1.11 Three-dimensional coordinate systems: **a** Cartesian right-handed, **b** cylindrical and **c** spherical coordinate systems

1.3.4   Cylindrical Coordinates

Extension of polar coordinates to three-dimensional space creates cylindrical and spherical coordinate systems. The cylindrical coordinates (Fig. 1.11b) can be thought of as polar coordinate planes displaced in the third dimension orthogonally to the original polar coordinate plane. Therefore, in addition to polar r and θ, the third coordinate h is added, which may have positive and negative values. The same ambiguity of the polar coordinates can create interesting shapes, and the same restricted domains for the values of r and θ can be set. The name "*cylindrical*" comes from the fact that by fixing a certain value of r and by allowing θ and h to take any values, we define the surface of an infinitely long cylinder. The cylindrical coordinates are therefore good for defining shapes which are topologically equivalent to cylinders, i.e. those shapes which surfaces we can mentally deform so that they become cylindrical surfaces.

With reference to Fig. 1.12, we can convert the cylindrical coordinates r, α, h to Cartesian coordinates as follows:

$$x = r \cos \alpha$$
$$y = h \tag{1.4}$$
$$z = -r \sin \alpha$$

We can immediately write that y-coordinate is equal to h. Next, x- and z-coordinates can be computed based on the right triangle rule, as in Fig. 1.8. However, here we put minus in front of the z-coordinate to reflect that according to the used layout of the coordinate systems we have to obtain negative z-coordinates within the first 180° of the cylindrical coordinate. The negative sign will be then corrected by the negative values of the *sin* function when the angle is greater than 180°.

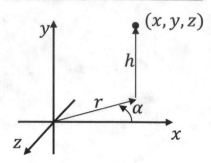

1.3.5   Spherical Coordinates

A different extension of polar coordinates to 3D space creates spherical coordinate
systems. Here, instead of one polar angle θ, two angles θ (called "zenith") and φ
called "azimuth" are used (Fig. 1.11c). These angles and the distance r to any point
then will uniquely define location of points in three-dimensional space if we restrict
the angular values so that one changes within π interval while another one—within
2π. Thus, by fixing the value of r and by incrementing one angle within 2π interval
we will define a circle, while by incrementing another angle within π interval we
will rotate this circle and thus create a sphere, hence the name of the coordinate
system. It is very efficient for defining location of point on surfaces which can be
deformed into the shape of sphere. In fact, this is a very frequently used coordinate
system since it is the very same geographical coordinate system with latitude and
longitude angles that is also used in Geographical Positioning System (GPS) in
every mobile device.

The conversion of spherical coordinates to Cartesian is performed following the
same rules as in Fig. 1.8. With reference to Fig. 1.13, the spherical coordinates r, α,
β will be converted to Cartesian coordinates as follows

$$x = r \sin \alpha \cos \beta$$
$$y = r \cos \alpha \tag{1.5}$$
$$z = r \sin \alpha \cos \beta$$

1.3.6   Visualization Pipeline

Let's look at the visualization pipeline backwards, from the image on the monitor
back to the object that has to be drawn.

Each pixel on the screen has its coordinates, which are integer numbers defined
in so-called *Device Coordinate System* (DCS). This is a two-dimensional Cartesian
Coordinate System with the origin in one of the screen corners and two coordinate
axes which use integer coordinates. Pixel coordinates define its address in the frame
buffer. Being converted into voltage potentials with a *Digital to Analog Converter*

Fig. 1.13 Conversion of spherical coordinates to Cartesian

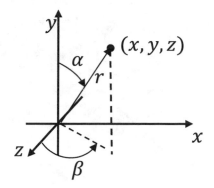

(DAC), these coordinates define the physical location of the pixel on the monitor. The color information controls the voltage applied to the electronics responsible for producing the color at the pixel location.

The coordinate range of the objects to be visualized is usually not constrained to the raster or physical size limits of the graphics device. Therefore, we introduce a so-called *World Coordinate System* (WCS)—the coordinate system that will be used in the application problem for defining location and geometry of the shapes, which then are to be displayed on the graphics device in its DCS. The WCS is usually a two- or three-dimensional Cartesian coordinate system. To define the WCS, we have to select where its origin will be locates, how the coordinate axes will be directed (i.e. right- or left-handed coordinate system), and what units of measurement will be used. The units of measurement for the WCS's coordinates depend on the application problem, and, generally, they are floating-point numbers. They can be angstroms for molecular modeling, meters for interior design, kilometers for virtual environments, parsecs for virtual stellar systems, etc. (Fig. 1.14).

Besides using the WCS, many *Modeling Coordinate Systems* (MCS) may be introduced as well within the WCS and used for modeling different shapes and groups of shapes. So a hierarchy of the MCSs can be created with one, and only one, WCS.

Suppose now that we have geometric models of our shapes defined in the WCS with their coordinates. The WCS is infinite while the DCS is finite. As such, we still need to introduce one more coordinate system to build a bridge from the WCS to the DCS. This is the *Viewer's Coordinate System* (VCS). Imagine that we are able to look at our models through the viewfinder of a camera. In that case, the coordinate system associated with this camera will be called VCS. It is defined by a position of the camera in the WCS as well as a point, which the camera looks at, as well as up and right vectors to define the camera tilt (Fig. 1.15).

The coordinate transformations, which are to be performed, are coordinate mappings from the WCS to the VCS, then from the three-dimensional VCS the coordinates which are to be projected onto a plane, and then a part of this infinite plane is to be converted into device coordinates of the DCS. To be able to define 2D coordinates in the projection plane which will be converted to DCS, we use a

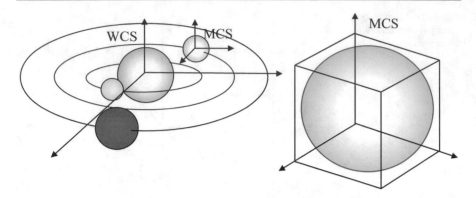

Fig. 1.14 Hierarchy of the world and modeling coordinate systems

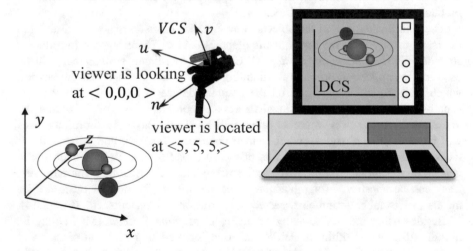

Fig. 1.15 Transformation from the world coordinates, through the viewer's coordinates to the device coordinates

concept of *Window* that is a rectangular area through which the viewers look at the WCS. It is like a camera's viewfinder. Whatever is visible through the window, will be then mapped into another rectangular area defined in DCS. This area is called *Viewport* (Fig. 1.16).

Do not mix up a window in the VCS with the window, which is a part of a graphical-user interface of the operating system. In fact, the graphics area inside the window on the screen is a viewport for the graphics application, which created this graphics window. This confusion with the terms was created historically, and in this book we refer to the term *window* as a rectangular area in the VCS, through which we look at the models to be displayed. The whole visualization pipeline can now be illustrated as it is shown in Fig. 1.17.

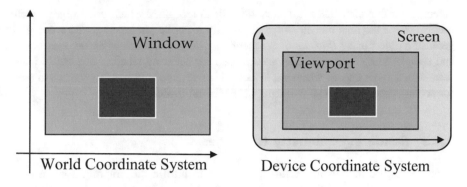

Fig. 1.16 Transformation from the window in the WCS to the viewport in the DCS

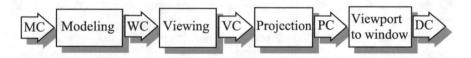

Fig. 1.17 The visualization pipeline

1.4 Geometric Algebra

1.4.1 Geometric Modeling

After the WCS has been defined, we have to create models of geometric shapes which will be eventually rendered to an image on the screen. Generally, when *geometric models* are being formed, they do not have neither color, texture, nor any other graphics attributes. They are to be added when graphics rendering is performed. *Geometric modeling* aims to create a model describing only geometry: size, location and topology of the real or imaginary shapes. When creating geometric models, we often work as if we are solving an algebraic problem, but instead of constants, variables, operations and relations, we introduce geometric entities such as *basic geometric objects* or *primitives*, *geometric operations* and *geometric relations*. A typical set of 3D geometric primitives comprises sphere, cylinder, cone, box, plane, polygonal shape. There is no such thing as one standard set of primitives, and if a set of primitives had to be defined, it would be introduced depending on the problem which has to be solved. Anyway, different graphics software tools usually offer an extensive set of basic objects to select from, with a possibility of adding more basic objects. These basic primitives can be used as building blocks for creating more complex geometric models. For doing this, geometric operations and relations are available. *Geometric operations* usually include linear transformations of coordinates as well as more sophisticated non-linear transformations. Linear transformations are also called *affine transformations* and they include translation, rotation and scaling. Non-linear transformations allow for achieving distortions of objects such that sometimes it is very difficult to recognize which basic objects were used to create the final model. Different primitives can be put together into a complex geometric object using *Boolean* or *Set-theoretic Operations*, such as *union*, *subtraction* and *complement*. *Geometric relations* are used for detecting whether one object intersects with another one, or whether it is located inside another object. They are also used for solving other topological problems of geometric modeling.

1.4.2 Rendering Geometry to Images

After geometric models are defined, we have to inform the graphics system how exactly to render them. For 2D geometric models, we have to define a *line type*, *width*, *color* and *filling style*, while for 3D shapes we define *material properties* and *transparency* of the objects. After that, provided we need a 2D image rather than a 3D physical model to be made, the viewing transformation, including WCS to VCS coordinate transformation, followed by projection and window-to-viewport transformations are to be performed, eventually leaving us with 2D coordinates of *graphics primitives* in the DCS. These graphics primitives are usually, point, straight lines and shaded polygons. Should we be developing an interactive application with the graphics interactive devices, we will then be able to locate individual points, primitives, or a group of primitives on the screen. By performing an inverse coordinate transformation from DCS back to the WCS, we will be able to change the original geometric models and see the results of these changes when they are rendered back to the graphics display.

1.4.3 Mathematical Functions in Geometric Modeling

We will define shapes and their transformations by using functions and indication how to sample them by the respective rendering algorithm. A function is a relation that uniquely associates numbers of one set with numbers of another set. A function is therefore a one-to-one or many-to-one relation. The set of possible input values of the function is called its *domain*, while the set of values that the function can produce is called its *range*. There are three ways to define functions analytically: *explicitly*, *implicitly* and *parametrically*. In either case, we have to use some coordinate system to produce coordinates.

In terms of coordinate transformations, *explicit representation* of one coordinate as a function of another one can be defined as

$$y = f(x) \text{ or } x = f(y) \text{ or } z = f(x, y) \tag{1.6}$$

Here, for every member of the function domain a valid value is computed in its range so that each function computation will produce valid coordinates of a point on the shape. Therefore, explicit functions are expected to be very efficient for rendering in terms of computational time. However, there can be some complications. Sometimes, the term "function" is also used to produce multiple values in the range. These functions are called *multivalued functions* or *multiple-valued functions*. For example, a square root function produces positive and negative values for any input, e.g. $\sqrt{4} = \pm 2$, while the inverse sine function produces infinite number of valid values for any input, e.g. $\arcsin(0) = n\,\pi$, where n is any integer number, positive or negative. The use of these functions requires additional considerations in visualization algorithms since usually still only one value will be produced by such functions implemented in mathematical libraries of programing languages, e.g. only positive values of the square root and so-called main values for inverse trigonometric functions will be computed.

Sometimes, we either cannot express one coordinate as a function of another, or purposely do not want to do it. In that case, we may use *implicit representation* of coordinate functions which is

$$f(x,y) = 0 \text{ or } f(x,y,z) = 0 \tag{1.7}$$

Here, for every member of the function domain a function value is computed and compared with 0. The function holds if its value is 0. Therefore, in case of implicit functions multiple computations will be required to understand which combinations of the coordinates from the function domain will satisfy the function. This type of functions will require much longer time for rendering but, as we will see, has other advantages over the explicit functions that allow implicit functions to be very much in demand in geometric modeling.

The third possible mathematical way of defining functions is *parametric*, when both coordinates are defined as independent functions of one or two parameters, depending on the shape we are modeling.

$$
\begin{aligned}
x &= x(u) & x &= x(u, v) \\
y &= y(u) & y &= y(u, v) \\
u &\in [u_1, u_2] & u &\in [u_1, u_2] \quad v \in [v_1, v_2]
\end{aligned}
\tag{1.8}
$$

Since parametric functions are explicit functions of parameters, they have all the computational advantages of explicit functions, i.e. they are fast for rendering tasks. We will also see that they are free from certain disadvantages of explicit functions because each coordinate is computed independently of another.

All three function representations are used in geometric modeling therefore there should be a way of converting between them. These ways of conversions are studied in mathematics.

To convert between an explicit function to implicit function, we have to move whatever was on the right of the equation to the left with a minus if front of it.

$$y = f(x) \Rightarrow y - f(x) = 0 \tag{1.9}$$

However, conversion from implicit to explicit function may be sometimes difficult or even impossible. Also, sometimes after the conversion multivalued function can be produced, as in this example:

$$x^2 + y^2 - 1 = 0 \Rightarrow y = \pm\sqrt{1 - x^2} \tag{1.10}$$

Conversion to parametric function is called parameterization. It can be performed as simple as declaring one of the variables a parameter, as in the following example:

$$y = f(x) \Rightarrow x = u \quad y = f(u) \tag{1.11}$$

However, as we will see, another and often more efficient solution can be found if we consider conversion between different coordinate systems so that polar, spherical or cylindrical coordinates will actually become parameters for representing Cartesian coordinates as it was illustrated in Fig. 1.8.

Inverse conversion from parametric to explicit and implicit functions has to be done by algebraic manipulations with the parametric equations (raising to power, multiplications, additions, subtractions, divisions, etc.) with the aim to eliminate the parameters, as in the following example:

$$
\begin{array}{c}
x = \sin^2(u) \\
+ \\
y = \cos^2(u) \\
\hline
x + y = \sin^2(u) + \cos^2(u) = 1 \ \Rightarrow \ x + y - 1 = 0
\end{array}
\tag{1.12}
$$

Also, the parameters can be eliminated by expressing them as a function of other variables and then by substitution to other parametric equations as in the following example:

$$
\begin{aligned}
x &= 2 + 3u \\
y &= 3 + u \Rightarrow u = y - 3 \Rightarrow x - 2 - 3(y-3) = x - 3y + 7 = 0
\end{aligned}
\tag{1.13}
$$

Note that during all these conversions between different types of functions the important matter is to preserve the domain of the variables. In some cases, the domain after the conversion cannot be expressed in a simple min–max form.

1.5   Summary

- Visualization replace real 3D light reflection/refraction with displaying the resulting images on some media.
- Random- and raster-scan display technologies are used in computer visualization.
- Invention of coordinate systems revolutionized geometry and created many new sciences including computer graphics and visualization.

- Computer graphics concerns making images with a computer.
- Visualization considers how to form an image from any data which even do not have an apparent geometric meaning.
- Cartesian, polar, cylindrical and spherical coordinate systems are mostly used in computer graphics for the purposes of defining geometric shapes mathematically.
- Visualization requires to define some mathematical model to be rendered into images.
- Shapes consist of geometry, colors, image textures, and geometrical textures.
- All shape components can be defined in their own coordinate systems and merged together into one object.
- Shapes can be further transformed and eventually grouped into one application coordinate system.
- Viewer and light sources have to be defined to render the scene.

References

1. How Vision Works, https://health.howstuffworks.com/mental-health/human-nature/perception/eye.htm
2. The Framing Game, https://www.vision3d.com/frame.html
3. Handbook of Visual Display Technology, ed. Janglin Chen, Wayne Cranton, Mark Fihn. Springer-Verlag GmbH Germany, https://doi.org/10.1007/978-3-642-35947-7, (2016)
4. How Computer Monitor Works, https://computer.howstuffworks.com/monitor5.htm
5. Stillwell J. Greek Geometry. In: Mathematics and Its History. Undergraduate Texts in Mathematics. Springer, New York, NY. https://doi.org/10.1007/978-1-4419-6053-5_2, (2010)
6. Roman Numerals, https://www.britannica.com/topic/Roman-numeral
7. Padmanabhan, T., Padmanabhan, V. The Dawn of Science. Glimpses from History for the Curious Mind, Springer, 2019.
8. Foley, J.D., van Dam, A., Feiner S.K., Hughes, J.F., Computer Graphics. Principles and Practice, Addison-Wesley, 1995.
9. Gonzalez, R.C., Woods, R.E., Digital Image Processing, Addison Wesley, 1993.
10. Spence, R., Information Visualization, Addison-Wesley, 2001.
11. Angel, E., Shreiner, D., Interactive Computer Graphics. A Top-Down Approach with WebGL, Pearson, 2015.
12. Introduction to Implicit Surfaces, /ed. Bloomenthal, J., Morgan Kaufmann Publishers, 1997.
13. Thalmann, N.M., Thalmann, D.,Computer Animation. Theory and Practice, Springer, 1990.
14. Watt, A. and Watt, M., Advanced Animation and Rendering Techniques. Theory and Practice, Addison-Wesley, 1992.
15. Watt, A., Policarpo, F., 3D Games. Animation and Advanced Real-time Rendering, vol. 2, Addison-Wesley, 2003.
16. Vince, J., Virtual Reality systems, Addison-Wesley, 1995.
17. Slater, M., Steed, A., Chrysanthou, Y., Computer Graphics and Virtual Environments. From Realism to Real-Time, Addison-Wesley, 2002.

Geometric Shapes

2

2.1 Sampling Geometry

Usually, behind images on the screen, we have *geometric models* of the displayed objects. These models define only geometric properties of the objects—their shape, size, orientation, and location in space. Geometric models do not have any *graphics attributes*, such as color, transparency, texture, etc. These attributes are added to geometry when it is to be visualized.

We will use the terms *geometric shapes* and *geometric objects* to refer to two- and three-dimensional points, curves, surfaces, or solids. Starting with definitions of a few simple shapes, we will learn how to extend these mathematical considerations to making any complex shapes.

Since coordinate systems define the location of points, we can classify shapes in two- and three-dimensional space by the *degree-of-freedom* (DOF) which the point may have while moving by the shape or within it (Fig. 2.1). Thus, with zero DOF, where the point cannot move anywhere at all, we can define points themselves. One DOF allows the point to move along a curve—forward and backward. Two DOF permit the point to move by a surface as a combination of forward–backward and sideway motions. Finally, three DOF, where motion inside and outside of the shape is added to the two DOF of surfaces, permit for defining a solid object.

Thinking of points, curves, surfaces, and solids as sets of all possible locations of points on them allows us to naturally find computer representation for these objects since we digitally sample space with computers. Then, we have to study which respective algorithmic procedures have to be used to define these points in the fastest and most precise way. Since these are often conflicting requirements, we will look at the most suitable mathematical ways to solve the problem for each class of objects.

© The Author(s), under exclusive license to Springer Nature Switzerland AG 2021
A. Sourin, *Making Images with Mathematics*, Undergraduate Topics
in Computer Science, https://doi.org/10.1007/978-3-030-69835-5_2

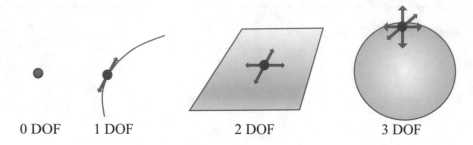

0 DOF   1 DOF   2 DOF   3 DOF

Fig. 2.1 Degrees-of-freedom when sampling shapes by points

2.2 Points

Points are defined by their coordinates which can be Cartesian, polar, cylindrical, spherical, etc.

Points can be displayed as individual points (Fig. 2.2a), used as geometric reference points for other models (Fig. 2.2b), or used in so-called particle and point-rendering systems (Fig. 2.2c, d) where thousands or millions of colored points are displayed so densely that even gaps between the points cannot be seen.

Actually, coordinates of the points can be produced by sampling any function definitions of curves, surfaces, and solid objects, which we will learn in the subsequent sections.

2.3 Curves

We will think of curves as moving points that sample them. We will determine the location of points by their coordinates, which are in turn computed by various mathematical functions. Computing the values of functions with different sampling resolution may produce either very dense location of points, so that all of them will look like a smooth curve, or we may sample a much smaller number of points and interpolate the space between them with some other curves like straight line segments (linear interpolation) or pieces of cubic curves (spline interpolation).

Let's begin with a definition of a few two-dimensional curves and learn by these examples which mathematical representation is the best for defining the curves. Mathematically, there are three ways to define two-dimensional curves: by implicit, explicit, and parametric functions.

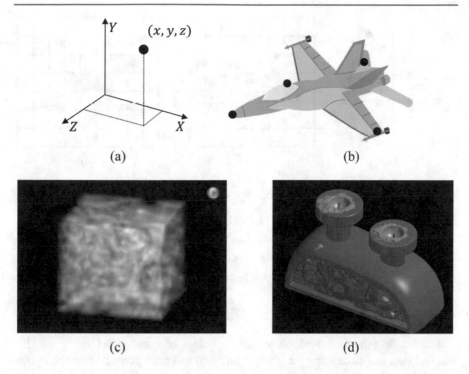

(a) (b)

(c) (d)

Fig. 2.2 Examples of using points: **a** Individual points; **b** Reference points of other models; **c** Point-based rendering of translucent gaseous shapes in mathematical simulations; **d** Photo-realistic rendering by displaying colored points without visible gaps between them

2.3.1 Straight Line

Straight line (Fig. 2.3a) can be defined by implicit function as

$$Ax + By + C = 0 \tag{2.1}$$

where the values of A, B, and C can be derived from the equation

$$\frac{y - y_1}{x - x_1} - \frac{y - y_2}{x - x_2} = 0 \tag{2.2}$$

which states that for any points P_1 and P_2 and any other point on the straight line with coordinates (x, y) the slope of the straight line remains the same.

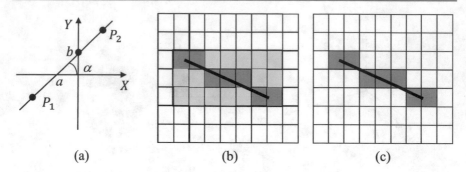

(a)   (b)   (c)

Fig. 2.3 **a** Straight line; **b** Drawing straight line on a raster monitor

The implicit equation of a straight line can be easily written as an *equation in intercepts* if we know the coordinates a and b of the points at which the straight line intersects the coordinate axes X and Y:

$$\frac{x}{a} + \frac{y}{b} = 1 \qquad \frac{x}{a} + \frac{y}{b} - 1 = 0 \qquad (2.3)$$

For drawing the line, we have to sample values of x and y and verify whether Eq. (2.1) holds. For defining the *straight line segment* from point P_1 to point P_2, the domains for $x-$ and $y-$ coordinates have to be $x \in [x_1, x_2], y \in [y_1, y_2]$, for a *ray cast* from P_1 through P_2: $x \in [x_1, \infty), y \in [y_1, \infty)$, and for the whole straight line: $x \in R, y \in R$, however, in place of an infinite domain in digital computers we have to define some very small and very large coordinates within the computer precision range. To display the straight line on a raster monitor, we have to sample integer values of pixel coordinates within the respective rectangular domain (Fig. 2.3b). However, this method of drawing will be slow since we have to sample lots of pixel coordinates that will not satisfy the implicit function Eq. (2.1).

The straight line can be also defined by explicit function

$$y = mx + b \qquad (2.4)$$

where b is the Y-axis intercept and m is a tangent of angle α. The value of m can be computed as

$$m = \frac{y - y_1}{x - x_1} \qquad (2.5)$$

In Eq. (2.4), we have to sample values of x to produce vales of y coordinates. Thus, for defining the straight line segment from point P_1 to point P_2, the domain for $x-$ coordinates has to be $x \in [x_1, x_2]$, for a ray cast from P_1 through P_2:

$x \in [x_1, \infty)$, and for the whole straight line: $x \in R$. Note that in case of the explicit function, every calculation will produce a valid pair of coordinates x and y. Then, to display the straight line on the raster monitor, we have to compute integer values of pixel coordinates (Fig. 2.3c) by incrementing the x−coordinate by 1 to compute the respective y−coordinate. This process is much faster than sampling all the pixels in the rectangular domain when the implicit function is used. However, there is a problem that for a vertical line we no longer can increment the x−coordinate since it remains the same for all y−coordinates. We then have to change the drawing algorithm that will compute different y−coordinates for the same x−coordinate. In other words, Eq. (2.4) does not work for the vertical lines, and therefore there is no single explicit function that can define a straight line with any slope.

We can overcome this deficiency by using a different type of explicit functions which are called parametric functions. Here, we can define the two-dimensional straight line with two explicit functions which independently produce x and y coordinates as functions of some other variable that we will call a parameter u:

$$x = x_1 + (x_2 - x_1)u$$
$$y = y_1 + (y_2 - y_1)u \qquad (2.6)$$

where x_1, y_1 and x_2, y_2 are coordinates of points P_1 and P_2 on the line, and $u \in [0, 1]$ for the straight line segment defined from P_1 and P_2,

$u \in [0, \infty)$ for the ray cast from P_1 through P_2, and.

$u \in R$ for the straight line passing through P_1 and P_2.

By changing the parameter u, we can illustrate how a point samples the curve having 1 DOF. When u increments from 0 to 1, the point moves from the starting point P_1 to the end point P_2 of the segment. By decrementing u from 1 to 0, while using the same Eq. (2.6), the coordinates of the point moving backwards from P_2 to P_1 will be produced. Moreover, the value of u reflects how far from the segment end points the currently computed point is located. Thus, $u = 0.25$ means that the respective point P with coordinates $x(u), y(u)$ will be located one quarter of the way from P_1 to P_2, $u = 0.5$ means halfway, while $u = 2$ means that the point will be located on the straight line outside the segment P_1, P_2 and at the double distance $\| P1\,P2\|$ from P_1. Refer to Fig. 2.4 for other examples of point calculations.

Parametric Eq. (2.6) can be used for drawing straight away. They will calculate coordinates of as many points on the straight line as needed depending on the increment of the parameter u. The value of the parameter increment may be adjusted to produce coordinates of the pixels. Note that Eq. (2.6) can be used for any orientation of the straight line while still retaining the fast computing performance of explicit functions.

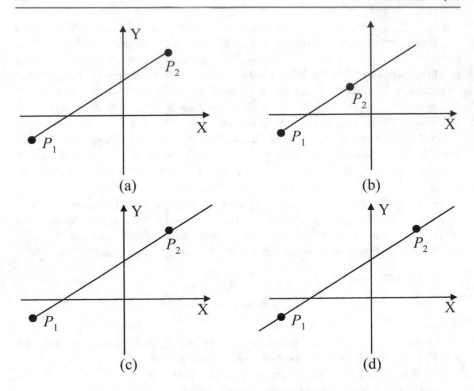

Fig. 2.4 Definition of straight line parametrically: **a** Straight line segment with $u \in [0, 1]$; **b** Straight line segment with $u \in [0, 2]$; **c** Ray with $u \in [0, \infty)$; **d** Straight line with $u \in R$.

2.3.2 Circle

Let's now look at the ways of defining a *circle*.

The origin-centered circle with radius r (Fig. 2.5a) can be defined by implicit functions in Cartesian coordinates by

$$r^2 - x^2 - y^2 = 0 \tag{2.7}$$

This equation originates from the Pythagoras theorem. For each right triangle inscribed into the circle in which the hypotenuse is the radius r, one adjacent side is $x-$coordinate and the other side is $y-$coordinate. To define a circle centered at any point with coordinates (x_0, y_0), the Eq. (2.7) has to be modified by including the inverse translation from the center point (x_0, y_0) to the origin:

$$r^2 - (x - x_0)^2 - (y - y_0)^2 = 0 \tag{2.8}$$

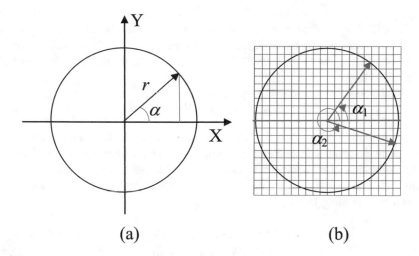

Fig. 2.5 **a** A circle defined in Cartesian and polar coordinates; **b** Sampling of the circle and its arc on the raster monitor

This equation can be used for drawing a circle by computing individual points on it or by sampling pixels on the raster monitor (Fig. 2.5b) within the square coordinate domains $x \in [x_0 - r, x_0 + r]$ and $y \in [y_0 - r, y_0 + r]$, however, like in the straight line case, it is a slow computational process since most of the points, while not belonging to the circle, still have to be sampled by computing Eq. (2.8). Also, for drawing a circular arc, it often becomes impossible to define a rectangular coordinate domain with min/max values of the $x-$ and $y-$ coordinates. For that rather a polar angle $\alpha \in [\alpha_1, \alpha_2]$ has to be used but it is not a coordinate in the Cartesian system.

Explicit equation of the circle can be derived from Eq. (2.8):

$$y = \pm\sqrt{r^2 - (x - x_0)^2} + y_0 \qquad (2.9)$$

However, we can see that this is a double-valued function: the positive values of the square root define the upper half of the circle while the negative values—the lower part. Like in the straight line case, we do not have one function defining any circle. That means that the drawing algorithm will have brunches while we expect to find single functions that are good for any curve.

Let's then explore parametric definitions of the circle. A simple parameterization can be done by declaring the $x-$ coordinate a parameter:

$$x = u$$
$$y = \pm\sqrt{r^2 - (u - x_0)^2} + y_0 \qquad (2.10)$$

However, the function still has the double-valued square root and we, therefore, have to look for other ways of creating parametric functions. Very often, function definitions made in other coordinate systems may help. Let's, therefore, look at polar coordinates. An origin-centered circle with radius R will be defined in polar coordinates as

$$r = R, \quad \alpha \in [0, 2\pi] \tag{2.11}$$

This definition means that for any polar angle α within 2π domain, we select a point at the distance R from the origin, which is actually a definition of a circle as a shape consisting of all points in a plane that are located at a given distance from a point called the center. Conversion from polar to Cartesian coordinates can be performed according to Fig. 1.8:

$$
\begin{aligned}
x &= r \cos \alpha = R \, \cos \alpha \\
y &= r \, \sin \alpha = R \, \sin \alpha \\
\alpha &\in [0, 2\pi]
\end{aligned}
\tag{2.12}
$$

These definitions allow for the computing of all the points on the circle by simply incrementing the value of α within 2π domain. Moreover, by incrementing α within any other domain $\alpha \in [\alpha_1, \alpha_2]$ and *offsetting* x and y definitions by the circle center coordinates we can define any circle or *arc* while still retaining the fast computing performance of explicit functions:

$$
\begin{aligned}
x &= R \, \cos \alpha + x_0 \\
y &= R \, \sin \alpha + y_0 \\
\alpha &\in [\alpha_1, \alpha_2]
\end{aligned}
\tag{2.13}
$$

Furthermore, with reference to Fig. 2.6, we can see that sampling with an equal increment provides a non-equal spacing of the points in case of the explicit functions Eq. (2.9) and a uniform spacing in case of the parametric representation Eq. (2.13) which is good for linear interpolation of the circle.

Note that in parametric functions Eq. (2.13), we used the parameter which is in fact the polar angle, hence if we need to make an arc that starts at 45° angle, as measured from axis X, the initial value of the parametric domain has to be $\frac{\pi}{4}$. This is very convenient but may not be always so when we derive parametric functions for other curves.

Furthermore, there can be other considerations taken into account to come up with parametric functions, and they may differ in their computational performance and the way how regular the point sampling is. Let's consider the following functions:

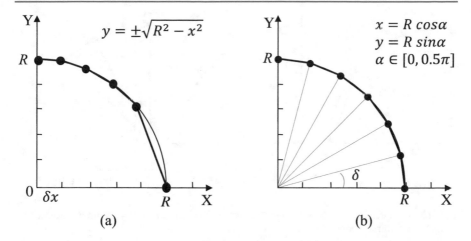

Fig. 2.6 Comparison of **a** explicit and **b** parametric functions when they are used for drawing a circle with a uniform increment of **a** δx of $x-$ coordinate and **b** $\delta \alpha$ of parameter α.

$$x = \frac{1 - u^2}{1 + u^2}, \quad y = \frac{2u}{1 + u^2} \quad u \in [0, 1] \tag{2.14}$$

They define a quarter of a circle with radius 1. This can be proved if we use the circle definition Eq. (2.7) and substitute in it the formulas for $x-$ and $y-$ coordinates:

$$1 - x^2 - y^2 = 1 - \left(\frac{1 - u^2}{1 + u^2}\right)^2 - \left(\frac{2u}{1 + u^2}\right)^2 = 1 - \frac{(1 + u^2)^2}{(1 + u^2)^2} = 0 \tag{2.15}$$

The circle is located in the first quarter of the coordinate plane because, according to Eq. (2.14), the $x-$ and $y-$ coordinates are always positive when the parameter values are obtained from $[0, 1]$. If we then use these representations for sampling points on the circle, we will see that the distribution of the points is better than what the explicit function Eq. (2.9) can produce but not as uniform as the previously considered parametric functions Eq. (2.13) can do (Fig. 2.7).

2.3.3 Ellipse

Let's now expand our knowledge to defining an *ellipse*.

The origin-centered ellipse with semi-axes a and b (Fig. 2.8a) can be defined by implicit functions in Cartesian coordinates as

$$1 - \left(\frac{x}{a}\right)^2 - \left(\frac{y}{b}\right)^2 = 0 \tag{2.16}$$

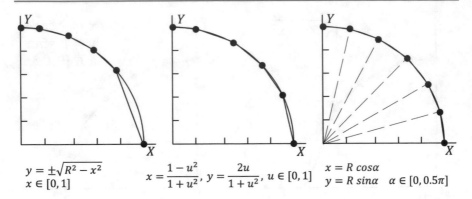

$$y = \pm\sqrt{R^2 - x^2}$$
$$x \in [0,1]$$

$$x = \frac{1 - u^2}{1 + u^2}, \, y = \frac{2u}{1 + u^2}, \, u \in [0,1]$$

$$x = R\cos\alpha$$
$$y = R\sin\alpha \quad \alpha \in [0, 0.5\pi]$$

Fig. 2.7 Comparison of explicit and two parametric functions when they are used for drawing a circle with uniform increments

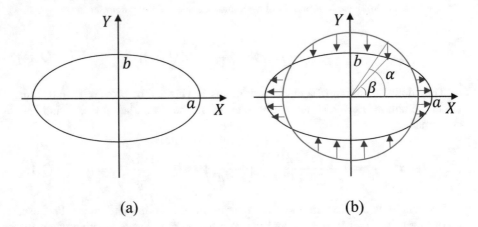

(a)   (b)

Fig. 2.8 a An ellipse defined in Cartesian coordinates; **b** Scaling a circle with radius 1 to the ellipse with semi-axes a an b

The ellipse centered at a point with coordinates (x_0, y_0) is then defined by

$$1 - \left(\frac{x - x_0}{a}\right)^2 - \left(\frac{y - y_0}{b}\right)^2 = 0 \tag{2.17}$$

When used for drawing, this equation has the same problems as the implicit equation of a circle Eq. (2.8): it is slow for sampling points on the ellipse and it cannot be used for defining elliptical arcs.

Explicit equation of the ellipse can be derived from Eq. (2.17) as

$$y = \pm b\sqrt{1 - \left(\frac{x - x_0}{a}\right)^2} + y_0 \tag{2.18}$$

Like the explicit equation of the circle Eq. (2.9), this is also a double-valued function: the positive values of the square root define the upper half of the ellipse while the negative values—the lower part. Like in the straight line and the circle cases, we do not have one function defining the whole ellipse, which would complicate drawing.

Parametric function for the ellipse can be obtained from the parametric functions of the circle if we think about the ellipse as the circle with radius 1 which was scaled in $x-$ and $y-$ directions to form the ellipse (Fig. 2.8b). Scaling can be done by multiplying by a and b the equations Eq. (2.13) defining a circle with radius 1:

$$\begin{aligned} x &= a \cdot 1 \cdot \cos\beta + x_0 = a\cos\beta + x_0 \\ y &= b \cdot 1 \cdot \sin\beta + y_0 = b\sin\beta + y_0 \\ \beta &\in [\beta_1, \beta_2] \end{aligned} \tag{2.19}$$

Note, however, that parameter β in these equations is no longer the polar angle α, as in the equations of the circle Eq. (2.13), since Eq. (2.19) reflects scaling of the circle which implies that the polar angle will change (Fig. 2.8). Parameter β equals to the angle α only for angles $0, \frac{\pi}{2}, \pi, \frac{3\pi}{2}, 2\pi$, etc. Therefore, to define an arc of the ellipse from a certain polar angle α we will need to calculate the respective parameter value β. It can be done by solving a set of simultaneous equations defining parametrically a straight line with a slope α and an implicit equation of the ellipse with semi-axes a and b to compute coordinates of the intersection point of a ray with the ellipse:

$$\begin{aligned} x &= u\cos\alpha \\ y &= u\sin\alpha \quad u \in [0, \infty) \\ 1 &- \left(\frac{x}{a}\right)^2 - \left(\frac{y}{b}\right)^2 = 0 \end{aligned} \tag{2.20}$$

By substituting the formulas for x and y into the third equation:

$$1 - \left(\frac{u\cos\alpha}{a}\right)^2 - \left(\frac{u\sin\alpha}{b}\right)^2 = 0 \tag{2.21}$$

we will solve it in terms of u. There will be two solutions since this is a quadric equation, and we have to select the one with a positive value since, according to Eq. (2.20), $u \in [0, \infty)$. After that, for the calculated value of x and y, we solve the parametric Eq. (2.19) in terms of the parameter β.

For example (see Fig. 2.9), for the ellipse with semi-axes 1.0 and 0.5 and
$\alpha_1 = \frac{\pi}{4}, \alpha_2 = \pi$:

$$\sin\tfrac{\pi}{4} = \cos\tfrac{\pi}{4} \approx 0.7$$
$$1 - \left(\tfrac{u \cdot 0.7}{1}\right)^2 - \left(\tfrac{u \cdot 0.7}{0.5}\right)^2 = 1 - 0.49u^2 - 1.96u^2 = 1 - 2.45u^2 = 0$$
$$u_1 = 0.64 \quad u_2 = -0.64$$
$$x = u\cos\alpha = 0.64 \cdot 0.7 = 0.448 \tag{2.22}$$
$$0.448 = 1 \cdot \cos\beta$$
$$\beta = \arccos 0.448 = 1.1\, rad = 63°$$

Though we studied only 3 curves—straight line, circle, and ellipse—the formulas used for their definition can be used for creating many more curves. Indeed,
the linear equation defining the straight line can be used for interpolating displacement of a point while it moves (translates) from one point to another, or for
changes of a size of a curve (e.g., changes of its length or scaling coefficient).
Equations defining a circle actually give us a way how to mathematically define the
rotation of a point about another point. Let's put these simple formulas together and
see how new curves can be easily created by parametric functions controlled by one
parameter $u \in [0, 1]$.

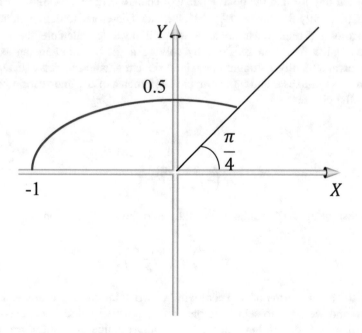

Fig. 2.9 An elliptical arc with the semi-axes $a = 1$ and $b = 0.5$ polar angles $\alpha_1 = \frac{\pi}{4}$ ($\beta_1 = 1.1$)
and $\alpha_2 = \pi(\beta_2 = \pi)$

2.3.4 Plethora of Curves

Starting with the parametric definition of the origin-centered circle with the constant radius a:

$$
\begin{aligned}
x &= a \cos(2\pi u) \\
y &= a \sin(2\pi u) \\
u &\in [0, 1]
\end{aligned}
\tag{2.23}
$$

we will first make the radius of the circle linearly increase as a function of the same parameter u:

$$
\begin{aligned}
x &= au \cdot \cos(2\pi u) \\
y &= au \cdot \sin(2\pi u) \\
u &\in [0, 1]
\end{aligned}
\tag{2.24}
$$

These formulas will define a curve called Archimedean spiral (Fig. 2.10a), which grows from the origin and spins about it eventually reaching the point with $x = a$.

We can change its rotation to the opposite clockwise direction, as shown in Fig. 2.10b, which is done by changing the sign of the argument in the trigonometric functions:

$$
\begin{aligned}
x &= au \cdot \cos(-2\pi u) \\
y &= au \cdot \sin(-2\pi u) \\
u &\in [0, 1]
\end{aligned}
\tag{2.25}
$$

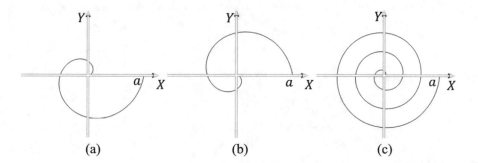

(a) (b) (c)

Fig. 2.10 Making the Archimedean spiral by parametric functions defining **a** counter-clockwise revolution, **b** clockwise revolution, and **c** three counter-clockwise revolutions

We can further change the number of rotations about the origin (see Fig. 2.10c) by adding a scaling factor ×3 to the angle of rotation, i.e. $3 \cdot 2\pi u = 6\pi u$:

$$
\begin{aligned}
x &= au \cdot cos(6\pi u) \\
y &= au \cdot sin(6\pi u) \\
u &\in [0, 1]
\end{aligned}
\tag{2.26}
$$

In this exercise, we created a new curve by combining linear functions defining the change of scale (radius and angle of rotation) with the functions defining the rotation of a point about the origin.

We can also experiment with different sampling values of the parameter domain followed by linear interpolation between the sampled points. The same equation of a circle Eq. (2.23) will eventually be used for the definition of very different curves if we use a smaller number of sampling points, as illustrated in Fig. 2.11. We can achieve an even more complicated look of the curves if we change the number of revolutions about the origin, as illustrated in Fig. 2.12.

All, these curves start on axis X at the coordinate $x = a$, however, it can be changed if we add an offset value δ to the angle of rotation.

$$
\begin{aligned}
x &= a \cdot cos(2\pi u + \delta) \\
y &= a \cdot sin(2\pi u + \delta) \\
u &\in [0, 1]
\end{aligned}
\tag{2.27}
$$

In Fig. 2.13, we rotate the curve counterclockwise by adding $\frac{\pi}{4}$ offset to the angle of rotation so that when $u = 0$ the starting angle becomes $\frac{\pi}{4}$.

Last but not least, let's experiment with the conversion of explicit function to a parametric and further linear and nonlinear transformation of it still using just a few simple formulas we have learnt so far.

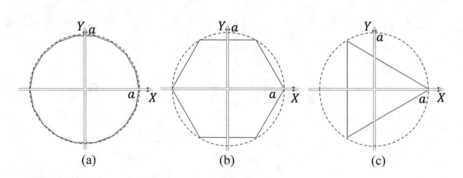

Fig. 2.11 Different curves obtained by a different sampling of the parameter u domain in the equations Eq. (2.13) which is done **a** 18 times, **b** 6 times, and **c** 3 times

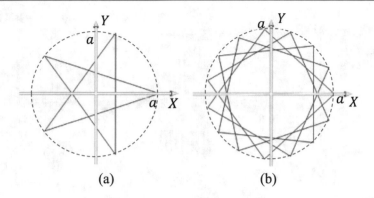

Fig. 2.12 Different curves obtained by a different sampling of the parameter u domain in Eq. (2.23) and a different number of rotations about the origin: **a** 5 samplings of u and $\times 2$ revolutions, i.e., $4\pi u$, and (b) 15 samplings of u and $\times 4$ revolutions, i.e.,$8\pi u$.

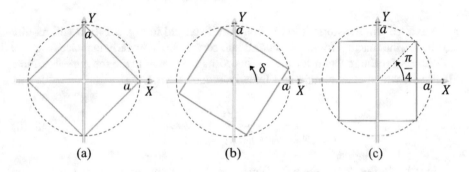

Fig. 2.13 Rotation of the square shape **a** obtained by 4 samplings of the parameter u domain in equation Eq. (2.13) by adding offset $\delta = \frac{\pi}{4}$ to the angle of rotation in **b** and **c**

Let's consider an explicit sine-wave function (Fig. 2.14a):

$$y = \sin(x) \quad x \in [-1, 0.6] \tag{2.28}$$

To convert this explicit function to parametric, we will do a simple parameterization by assigning variable x to be a parameter u:

$$
\begin{aligned}
x &= -1 + u(0.6 - (-1)) = -1 + 1.6u \\
y &= \sin(2\pi u) \\
u &\in [0, 1]
\end{aligned}
\tag{2.29}
$$

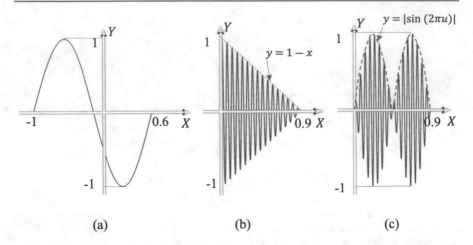

Fig. 2.14 Transformations of the sine-wave function

Next, we will transform it to the function displayed in Fig. 2.14b. It will require to change the function for the $x-$ coordinate so that it will linearly increase from 0.0 to 0.9. As for the function for the $y-$ coordinate, we will make the curve oscillate 16 times, while its amplitude will linearly reduce from 1 to 0:

$$x = 0.9u$$
$$y = (1 - u)\sin(16 \cdot 2\pi u) \qquad (2.30)$$
$$u \in [0, 1]$$

Finally, in Fig. 2.14c we will change the linear transformation of the function amplitude to the nonlinear function controlled by another *sin* function:

$$x = 0.9u$$
$$y = |\sin(2\pi u)|\sin(16 \cdot 2\pi u) \qquad (2.31)$$
$$u \in [0, 1]$$

2.3.5   Three-Dimensional Curves

For defining *3D curves*, we only can use parametric representations, because if we add one more coordinate to the explicit or implicit functions defining plane curves they will become functions defining surfaces, and 3D curves can only be defined as their intersections. On the other hand, parametric functions can be easily extended to any dimension just by adding the respective function defining the new dimension coordinate as another explicit function of the same parameter.

$$
\begin{aligned}
x &= x(u) \\
y &= y(u) \\
z &= z(u) \\
u &\in [u_1, u_2]
\end{aligned}
\tag{2.32}
$$

Thus, we can add one more linear equation to the parametric definition of the 2D straight line, and it will define a 3D straight line:

$$
\begin{aligned}
x &= x_1 + (x_2 - x_1)u \\
y &= y_1 + (y_2 - y_1)u \\
z &= z_1 + (z_2 - z_1)u
\end{aligned}
\tag{2.33}
$$

where x_1, y_1, z_1 and x_2, y_2, z_2 are coordinates of points P_1 and P_2 on the line, and $u \in [0, 1]$ for the straight line segment defined from P_1 and P_2,
$u \in [0, \infty)$ for the ray cast from P_1 through P_2, and.
$u \in R$ for the straight line passing through P_1 and P_2.

In a similar way, we can add the third coordinate equation to the equations of a 2D circle to place it in a plane parallel to the coordinate plane XY and with the center at coordinates x_0, y_0, z_0:

$$
\begin{aligned}
x &= R \cos(2\pi u) + x_0 \\
y &= R \sin(2\pi u) + y_0 \\
z &= z_0 \\
u &\in [0, 1]
\end{aligned}
\tag{2.34}
$$

By changing the constant $z-$ coordinate and by modifying the $x-$ and $y-$ functions we can easily create definitions of the curves displayed in Fig. 2.15. Specifically, the expanding *helical curve* in Fig. 2.15a is defined by

$$
\begin{aligned}
x &= 0.6u \cos(8\pi u) \\
y &= 0.6u \sin(8\pi u) \\
z &= -0.5 + 1.5u \\
u &\in [0, 1]
\end{aligned}
\tag{2.35}
$$

In Fig. 2.15b, the radius is fixed to define a cylindrical helical curve:

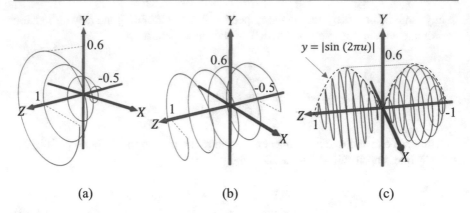

Fig. 2.15 3D curves created from the equation of a 2D circle

$$x = 0.6 \cos(8\pi u)$$
$$y = 0.6 \sin(8\pi u)$$
$$z = -0.5 + 1.5u \tag{2.36}$$
$$u \in [0, 1]$$

Finally in Fig. 2.15c, the radius is non-linearly scaled by sine-wave amplitude modulation:

$$x = 0.6 \, |\sin(2\pi u)| \, \cos(36\pi u)$$
$$y = 0.6 \, |\sin(2\pi u)| \, \sin(36\pi u)$$
$$z = -1 + 2u \tag{2.37}$$
$$u \in [0, 1]$$

2.4 Surfaces

We will think of surfaces as sets of points that sample surfaces by moving on them or as sets of points created by moving curves. When sampling curves, points may have only one degree of freedom to move—forward and backward. When sampling surfaces, points can also move sideways which adds one more degree of freedom. Surfaces can be then visualized either by sampling and displaying individual points (as in *ray tracing*), or by displaying polygons (usually, triangles) that interpolate surfaces.

Let's consider a few basic surfaces and explore how explicit, implicit, and parametric functions can be used for their definition.

2.4.1 Plane

A *plane* is defined by the implicit equation as

$$Ax + By + Cz + D = 0 \tag{2.38}$$

The values of the coefficients A, B, C, and D can be calculated by eliminating D (by dividing the equation by D) and by further solving a set of three plane equations using the coordinates for three non-collinear points on the plane:

$$\frac{A}{D}x + \frac{B}{D}y + \frac{C}{D}z + 1 = A'x + B'y + C'z + 1 = 0 \tag{2.39}$$

This method, based on using coordinates of three non-collinear points, can be then used for calculating the plane equation for all the possible geometric ways of defining planes such as by /1/ two intersecting lines, /2/ two parallel lines, /3/ a point and a line not containing this point. We just have to sample the lines to obtain coordinates of the three non-collinear points on the plane.

Alternatively, we can derive the plane equation based on the fact that for a normal vector $\mathbf{N} = [A\ B\ C]$, which is orthogonal to the plane (Fig. 2.16), a dot product with any other vector cast from a point $\mathbf{P}$ to any other point on the plane will always be equal to 0 because $cos\frac{\pi}{2} = 0$:

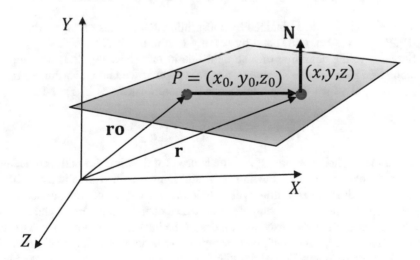

Fig. 2.16 Definition of a plane by its normal N and a point P on the plane

$$\mathbf{N} \cdot (\mathbf{r} - \mathbf{r}_0) = \|\mathbf{N}\| \|\mathbf{r} - \mathbf{r}_0\| \cos\frac{\pi}{2} = \|\mathbf{N}\| \|\mathbf{r} - \mathbf{r}_0\| 0 = 0$$

$$A(x - x_0) + B(y - y_0) + C(z - z_0) \qquad\qquad (2.40)$$

$$= Ax + By + Cz + (-Ax_0 - By_0 - Cz_0) = 0$$

Thus, we can conclude that the coefficients A, B, C are in fact coordinates of the vector orthogonal to the plane, called a plane normal vector (or just a *normal*). Hence the plane can be also defined by a point on the plane and coordinates of a normal to the plane. The normal vector can be computed as a cross product of two vectors on the plane that requires us to know the coordinates of three non-collinear points P_1, P_2, P_3 on the plane:

$$\mathbf{P}_1 = [x_1 \quad y_1 \quad z_1], \ \mathbf{P}_2 = [x_2 \quad y_2 \quad z_2], \ \mathbf{P}_3 = [x_3 \quad y_3 \quad z_3]$$

$$\mathbf{a} = \mathbf{P}_2 - \mathbf{P}_1 = [x_2 - x_1 \quad y_2 - y_1 \quad z_2 - z_1]$$

$$\mathbf{b} = \mathbf{P}_3 - \mathbf{P}_1 = [x_3 - x_1 \quad y_3 - y_1 \quad z_3 - z_1]$$

$$\mathbf{a} \times \mathbf{b} = \begin{bmatrix} \mathbf{i} & \mathbf{j} & \mathbf{k} \\ x_a & y_a & z_a \\ x_b & y_b & z_b \end{bmatrix} = \mathbf{i} \begin{Vmatrix} y_a & z_a \\ y_b & z_b \end{Vmatrix} - \mathbf{j} \begin{Vmatrix} x_a & z_a \\ x_b & z_b \end{Vmatrix} + \mathbf{k} \begin{Vmatrix} x_a & y_a \\ x_b & y_b \end{Vmatrix} \qquad (2.41)$$

$$= \mathbf{i}(y_a z_b - y_b z_a) + \mathbf{j}(x_b z_a - x_a z_b) + \mathbf{k}(x_a y_b - x_b y_a)$$

$$\mathbf{N} = [y_a z_b - y_b z_a \quad x_b z_a - x_a z_b \quad x_a y_b - x_b y_a] = [A \quad B \quad C]$$

The value of D is then obtained by substituting coordinates of any of the points P_1, P_2, P_3 into the Eq. (2.37) and by solving it in terms of D.

Similarly to the definition of a straight line in intercepts Eq. (2.3), the implicit equation of a plane can be also written in *intercepts* if we know coordinates a, b, c of the points at which the plane intersects the coordinate axes X, Y, and Z:

$$\frac{x}{a} + \frac{y}{b} + \frac{z}{c} = 1 \qquad \frac{x}{a} + \frac{y}{b} + \frac{z}{c} - 1 = 0 \qquad\qquad (2.42)$$

When the implicit equation of a plane is used for drawing, the same advantages and disadvantages, as for drawing a straight line defined implicitly, will be observed. The drawing algorithm will sample the $x-, y-, z-$ coordinate domain of the plane (bounding box) to compute coordinates of the points on the plane which are then either displayed or used as vertices of the interpolating polygons. This is a slow process since most of the points sampled in the domain will not belong to the plane. Any spatial orientation of the plane will be supported but we are not able to immediately use the plane equation for displaying any specific bounded part of it (e.g., triangular or four-sided quadrilateral polygon). The infinite plane will be then displayed as a finite polygon truncated by the geometric boundaries of the coordinate domain (bounding box) which will look differently depending on the plane

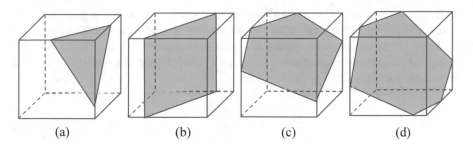

Fig. 2.17 Intersections of a plane with a boundary box: **a** triangle, **b** four-sided polygon, **c** five-sided polygon, **d** six-sided polygon

orientation and the domain location (three-, four-, five-, or six-sided polygon) as illustrated in Fig. 2.17.

The explicit equation of a plain can be derived for any of three coordinates from Eq. (2.38), e.g.:

$$z = \frac{-Ax - By - D}{C} \tag{2.43}$$

However, it will cause the same problem as any explicit equation: it will not be able to define any orientation of the plane. For example, Eq. (2.43) for the $z-$ coordinate does not exist for planes parallel to the Z-axis (Fig. 2.18) since it will become a multivalued function: for any pair of $x-$ and $y-$coordinates we will have infinite number of all possible $z-$ coordinates. It is not then a simple computation of one $z-$ coordinate for any given x and y, and it will require branching in the visualization algorithm that we want to avoid. Therefore, as in the case of curves,

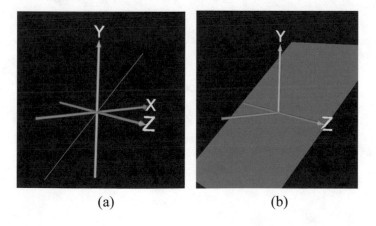

Fig. 2.18 The equation of the straight line **a** becomes the equation of a plane in 3D **b**

explicit functions are seldom used for visualization unless there is a guarantee that there will not be any special cases that have to be treated differently.

Let's, therefore, explore parametric representations. The 2 DOF of a point sampling the plane will be reflected by the two parameters that we have to use in the equations for x-, $y-$ and $z-$ coordinates:

$$\begin{aligned} x &= x(u, v) \\ y &= y(u, v) \\ z &= z(u, v) \\ u \in [u_1, u_1], \qquad &v \in [v_1, v_1] \end{aligned} \qquad (2.44)$$

We can then derive the parametric equations for a plane based on the understanding of Cartesian coordinates in a vector form (Fig. 2.19).

Each Cartesian axis X, Y, Z has a respective unit vector $\mathbf{i}, \mathbf{j}, \mathbf{k}$ associated with it. Then, any point with coordinates (x, y, z) is represented by its position vector which is obtained as a sum of three vectors $x\mathbf{i}$, $y\mathbf{j}$, and $z\mathbf{k}$. We will use the same vector definitions for a plane but in place of Cartesian unit vectors $\mathbf{i}, \mathbf{j}, \mathbf{k}$ and coordinates x, y, z we will use vectors defined by three points P_1, P_2, P_3 given on the plane and parameters u and v (Fig. 2.20).

Then, point P can be defined within the plane as a sum of vectors $[P_1 P_2]$ and $[P_1 P_3]$ which are scaled by parameters u and v:

$$\mathbf{P} = u(\mathbf{P}_2 - \mathbf{P}_1) + v(\mathbf{P}_3 - \mathbf{P}_1) \quad u, v \in R \qquad (2.45)$$

The position vector of point P in the Cartesian coordinate system XYZ is then obtained as

$$\mathbf{P} = \mathbf{P}_1 + u(\mathbf{P}_2 - \mathbf{P}_1) + v(\mathbf{P}_3 - \mathbf{P}_1) \quad u, v \in R \qquad (2.46)$$

Fig. 2.19 Vector representation of the Cartesian coordinates

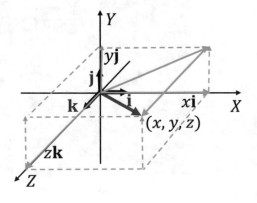

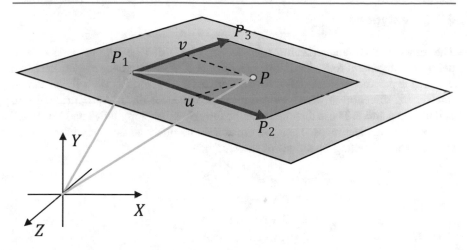

Fig. 2.20 Parameterization of a plane

In terms of Cartesian coordinates, the obtained parametric equations can be then written as

$$
\begin{aligned}
x &= x_1 + u(x_2 - x_1) + v(x_3 - x_1) \\
y &= y_1 + u(y_2 - y_1) + v(y_3 - y_1) \\
z &= z_1 + u(z_2 - z_1) + v(z_3 - z_1) \\
u, v &\in R
\end{aligned}
\tag{2.47}
$$

The obtained equations are linear, and we can recognize there two merged equations of straight lines from P_1 to P_2 and from P_1 to P_3:

$$
\text{Straight line from } P_1 \text{ to } P_2
$$
$$
\mathbf{P} = \underbrace{\mathbf{P_1} + u(\mathbf{P_2} - \mathbf{P_1})} + \underbrace{v(\mathbf{P_3} - \mathbf{P_1})} \quad u, v \in R
\tag{2.48}
$$
$$
\text{Straight line from } P_1 \text{ to } P_3
$$

When used for visualization, this parametric representation is fast for computation and capable of displaying planes with any orientation, however, the infinite plane will be displayed as a finite four-sided polygon constrained by the domains of u and v and always proportional to the shape of the parallelogram defined by the three points P_1, P_2, P_3, as displayed in Fig. 2.20. Thus, if the parameters are constrained by the domains $u, v \in [0, 1]$, then the formulas Eq. (2.47) will compute point P_1 for $u, v, = 0$; point P_2 for $u = 1, v, = 0$; and point P_3 for $u = 0, v, = 1$.

2.4.2 Polygons

The concept of parameterization used in Eq. (2.47) can be further expanded to definition of general four-sided polygons or *quads*. With reference to Fig. 2.21, we will define parametrically a polygon with vertices P_1, P_2, P_3, P_4. We will select two intersecting direction in which we can measure distances linearly proportionally to parameters u and v. These directions can be thought of as "longitude" and "latitude" used in geographical maps. Then, we are able to uniquely address any point within the polygon by setting up values of the parameters:

$$u = 0.0, v = 0.0 : \quad P_1$$
$$u = 1.0, v = 0.0 : \quad P_2$$
$$u = 0.0, v = 1.0 : \quad P_3$$
$$u = 1.0, v = 1.0 : \quad P_4$$
$$u = 0.5, v = 0.5 : \quad middle\,point$$

Let's define it in a vector form as we did it for the plane. For any point P' and point P'', which are located on the opposite sides of the polygon and controlled by the same value of parameter u, we can write

$$\mathbf{P}' = \mathbf{P}_1 + u(\mathbf{P}_2 - \mathbf{P}_1)$$
$$\mathbf{P}'' = \mathbf{P}_3 + u(\mathbf{P}_4 - \mathbf{P}_3)$$

(2.49)

Next, for any point P located on the segment joining points P' and P'' and controlled by parameter v we can write

$$\mathbf{P} = \mathbf{P}' + v(\mathbf{P}'' - \mathbf{P}')$$

(2.50)

By substituting into it Eq. (2.49), we obtain

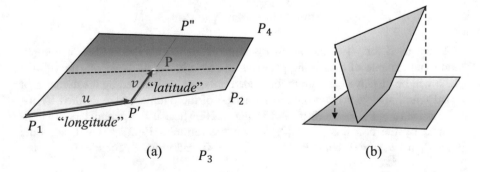

Fig. 2.21 Parameterization of four-sided polygon

$$\begin{aligned}
\mathbf{P} &= \mathbf{P}_1 + u(\mathbf{P}_2 - \mathbf{P}_1) + v(\mathbf{P}_3 + u(\mathbf{P}_4 - \mathbf{P}_3) - \mathbf{P}_1 + u(\mathbf{P}_2 - \mathbf{P}_1)) \\
&= \mathbf{P}_1 + u(\mathbf{P}_2 - \mathbf{P}_1) + v(\mathbf{P}_3 - \mathbf{P}_1 + u(\mathbf{P}_2 - \mathbf{P}_1 + \mathbf{P}_4 - \mathbf{P}_3)) \\
&= \mathbf{P}_1 + u(\mathbf{P}_2 - \mathbf{P}_1) + v(\mathbf{P}_3 - \mathbf{P}_1) + uv(\mathbf{P}_2 - \mathbf{P}_1 + \mathbf{P}_4 - \mathbf{P}_3) \\
u,&\, v, \in [0,1]
\end{aligned} \tag{2.51}$$

or in the Cartesian coordinates form:

$$
\overbrace{}^{\text{Linear part}} \qquad \overbrace{}^{\text{Non-linear part}}
$$

$$\begin{aligned}
x &= x_1 + u(x_2 - x_1) + v(x_3 - x_1) + uv(x_2 - x_1 + x_4 - x_3) \\
y &= y_1 + u(y_2 - y_1) + v(y_3 - y_1) + uv(y_2 - y_1 + y_4 - y_3) \\
z &= z_1 + u(z_2 - z_1) + v(z_3 - z_1) + uv(z_2 - z_1 + z_4 - z_3)
\end{aligned} \tag{2.52}$$

$$u, v, \in [0,1]$$

2.4.3 Bilinear Surfaces

Though linear equations Eq. (2.47) are a part of the definitions Eq. (2.52), they also have a nonlinear part with the uv term, i.e. Equation (2.52) are obtained by linear interpolation first in one direction and then again in the other direction, and though each step is linear, the interpolation as a whole is not linear but rather quadratic in any sampled location. These definitions are called *bilinear representations*. The polygons defined this way may not be only plain polygons but rather *patches* of 3D curved surfaces as illustrated in Fig. 2.21b.

Let's show what the obtained bilinear definitions can do.

In Fig. 2.22a–c, several bilinear surfaces are displayed. With the same definitions, they only differ in $y-$ coordinates of point P_1 in Fig. 2.22b, and points P_1 and P_4 in Fig. 2.22c. In Fig. 2.22d, the bi-linear nature of the surface is illustrated by the so-called *wireframe* drawing mode where we can see that the surface is actually defined by straight line segment moving so that its endpoints are moving by straight lines.

Bilinear representation can be also used for defining triangular polygons if we make equal any two of the neighboring vertices, as illustrated in Fig. 2.22e–f. The parametric definitions will then be simplified accordingly however will still remain nonlinear.

Note, however, that though the bilinear representations allow for defining finite plane polygons (quads or triangles) with the parameters within the domain [0, 1], they cannot be used for defining the whole plane by extending the domain. In contrast to the parametric plane definitions Eq. (2.47), setting the u, v domains outside the interval [0, 1] will not allow for defining a bigger polygon similar to the original with vertices P_1, P_2, P_3, P_4 since the parametric axes (directions in which the parameters u, v change) may eventually intersect, and it will lead to twisting the surface as it is illustrated in Fig. 2.23. In the case of a rectangular polygon (Fig. 2.23a) though, it

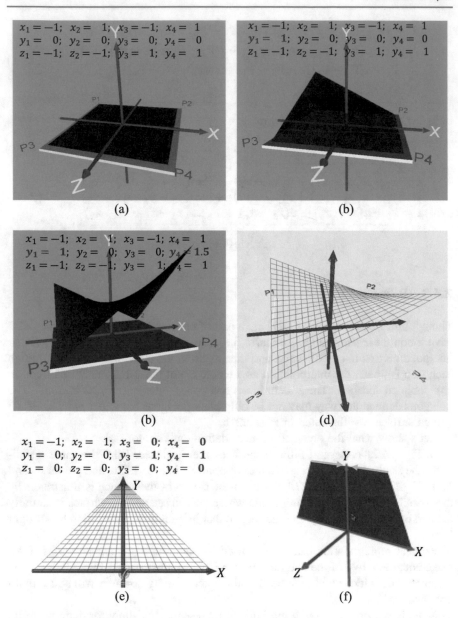

Fig. 2.22 a–c Examples of bilinear surfaces; **d** Wireframe rendering of bilinear surface; **e–f** Using bilinear surfaces for defining triangular polygons

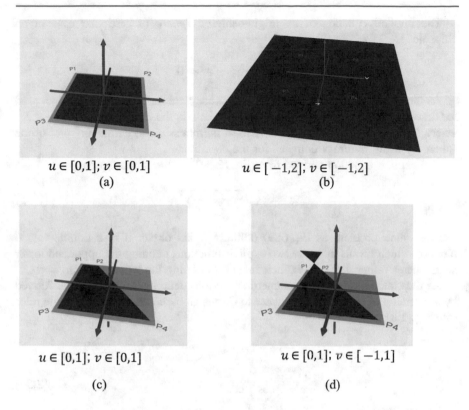

$u \in [0,1]; v \in [0,1]$
(a)

$u \in [-1,2]; v \in [-1,2]$
(b)

$u \in [0,1]; v \in [0,1]$

(c)

$u \in [0,1]; v \in [-1,1]$

(d)

Fig. 2.23 **a** Bilinear square polygon; **b** Bigger square polygon is displayed when the parameter domain expanded beyond [0, 1]; **c** Non-rectangular polygon; **d** The polygon twists when the parameter domain expands beyond [0, 1]

will work when the u, v domains increase and a bigger rectangular polygon will be displayed (Fig. 2.23b). However, if the original polygon is not rectangular, as in Fig. 2.23c, increasing the u, v domains outside the [0, 1] interval makes the polygon twist as it is shown in Fig. 2.23d.

2.4.4 Quadrics

The bilinear surface exhibits some properties of quadric surfaces which allow the surface to be curved. Let's consider now the quadric surfaces. The general quadric implicit equation defining all of them is

$$ax^2 + by^2 + cz^2 + dxy + eyz + fxz + gx + hy + kz + m = 0 \qquad (2.53)$$

Depending on the values of the coefficients, the equation defines *ellipsoid, elliptic paraboloid, hyperbolic paraboloid, hyperboloid of one sheet, hyperboloid of two sheets, cone, elliptic cylinder, hyperbolic cylinder, parabolic cylinder*.

The implicit equation of an origin-centered *sphere* (a particular case of an ellipsoid) is

$$R^2 - x^2 - y^2 - z^2 = 0 \tag{2.54}$$

As the implicit equation of a circle, it is rather slow for computations when used for drawing since we have to sample many points in the domains $x, y, z \in [-R, R]$, where R is the radius of the sphere. Also, it is impossible to define any part of the sphere by using min/max domains for the coordinates.

The explicit equation derived from Eq. (2.54) for the z-coordinate

$$z = \pm\sqrt{R^2 - x^2 - y^2} \tag{2.55}$$

has the same problem as Eq. (2.9) defining a 2D circle: it is a double-valued function which means there are two explicit functions defining the upper and lower hemispheres rather than one explicit function defining the whole sphere.

As illustrated in Fig. 2.24, parametric functions defining a sphere can be derived by conversion of spherical coordinates to Cartesian based on the right triangle rules (see Fig. 1.8):

$$\begin{aligned}
x &= r\cos\varphi\sin\theta \\
y &= r\sin\varphi \\
z &= r\cos\varphi\cos\theta \\
\varphi &\in \left[-\frac{\pi}{2}, \frac{\pi}{2}\right],\ \theta \in [-\pi, \pi]
\end{aligned} \tag{2.56}$$

Fig. 2.24 Conversion of spherical coordinates to Cartesian

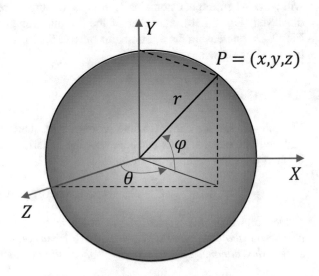

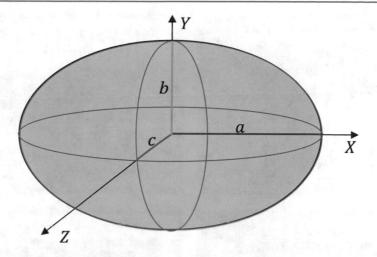

Fig. 2.25 Ellipsoid

Note that one parameter has a π domain, while another one—2π.

For the ellipsoid (Fig. 2.25), we can write the implicit, explicit, and parametric functions straight away, recalling how we did it for the 2D ellipse.

Implicit function:

$$1 - \left(\frac{x}{a}\right)^2 - \left(\frac{y}{b}\right)^2 - \left(\frac{z}{c}\right)^2 = 0 \tag{2.57}$$

Explicit function:

$$z = \pm c \sqrt{1 - \left(\frac{x}{a}\right)^2 - \left(\frac{y}{a}\right)^2} \tag{2.58}$$

Parametric functions:

$$\begin{aligned}
x &= a \cos\varphi \sin\theta \\
y &= b \sin\varphi \\
z &= c \cos\varphi \cos\theta \\
\varphi &\in \left[-\frac{\pi}{2}, \frac{\pi}{2}\right], \theta \in [-\pi, \pi]
\end{aligned} \tag{2.59}$$

Let's experiment with spheres and ellipsoids defined implicitly. To define a sphere centered at a point with coordinates x_0, y_0, z_0, we have to offset the coordinates of its center (Fig. 2.26a). An ellipsoid with semi-axes 1.1, 0.5, and 0.75 is defined in Fig. 2.24b. In Fig. 2.26(d, e) we define the so-called "super-ellipsoids"

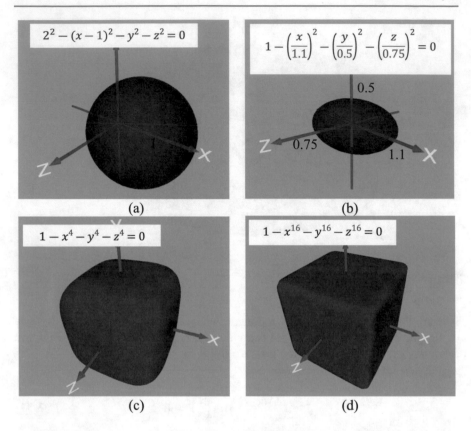

$$2^2 - (x-1)^2 - y^2 - z^2 = 0$$

$$1 - \left(\frac{x}{1.1}\right)^2 - \left(\frac{y}{0.5}\right)^2 - \left(\frac{z}{0.75}\right)^2 = 0$$

(a)   (b)

$$1 - x^4 - y^4 - z^4 = 0$$

$$1 - x^{16} - y^{16} - z^{16} = 0$$

(c)   (d)

Fig. 2.26 Experiments with spheres and ellipsoids defined implicitly

by changing the exponent 2 to 4 and 16, respectively, which changes a sphere into nearly a cube.

Two more quadric surfaces—a cone and a cylinder—are defined implicitly in Fig. 2.27. The cone is a symmetrical object consisting of two parts. The cylinder definition is in fact a definition of a plane ellipse in coordinates x and y. It becomes a definition of a cylinder in 3D space for any $z-$ coordinate by infinite displacing of the ellipse along the Z-axis.

In Fig. 2.28, we experiment with parametric definitions of a sphere. The sphere with radius 0.7 is centered at coordinates 0.9, 0, 0 in Fig. 2.28a. By changing the domain of parameter u from 2π to π, we display half a sphere in Fig. 2.28b.

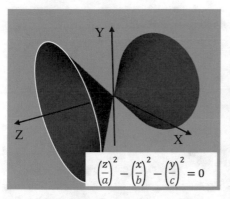
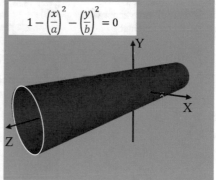

Fig. 2.27 A cone and a cylinder defined implicitly

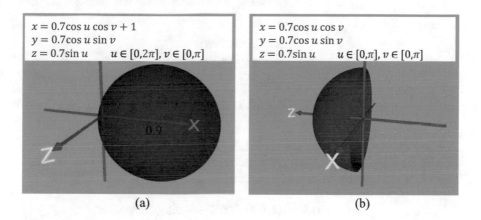

Fig. 2.28 Experiments with spheres defined parametrically

2.4.5 Making Surfaces by Sweeping Curves

In the same way, how we created new curves from only a few formulas defining linear interpolation and rotation, we now can create new surfaces by parametric formulas based on our knowledge of how to define planes, bilinear surfaces, and quadrics.

Let's define a cylinder displayed in Fig. 2.29 parametrically. We will consider two parameters as two degrees of freedom which can be used by a point traveling on the cylinder's surface. The cylinder can be thought of as a circle which is then translated along the cylinders' axis. An origin-centered circle with radius 2 placed in the coordinate plane XY is defined by

Fig. 2.29 Parametric
definition of the cylinder with
radius 2 and height 1.5

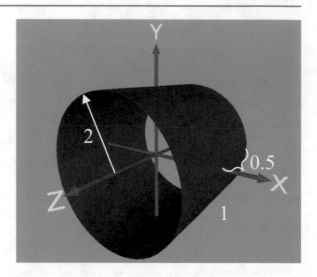

$$x = 2\cos(2\pi u)$$
$$y = 2\sin(2\pi u)$$
$$z = 0 \qquad\qquad\qquad (2.60)$$
$$u \in [0, 1]$$

While the points sampling the circle are always located in the coordinate plane
XY with the 0 $z-$ coordinate, the $z-$ coordinates of the cylinder have to span from $-$
0.5 to 1, which can be defined as a parametric function of another parameter v:

$$z = -0.5 + v(1 - (-0.5)) = -0.5 + 1.5v$$
$$v \in [0, 1] \qquad\qquad (2.61)$$

Next, we will put Eq. (2.60) and Eq. (2.61) together by replacing $z = 0$ in the
Eq. (2.60) with the Eq. (2.61) to obtain the parametric formulas of the cylinder:

$$x = 2\cos(2\pi u)$$
$$y = 2\sin(2\pi u)$$
$$z = -0.5 + 1.5v \qquad\qquad (2.62)$$
$$u, v \in [0, 1]$$

In the similar way, we can define a sphere as a circle defined in XY-plane which
rotates by angle π about axis X (Fig. 2.30).

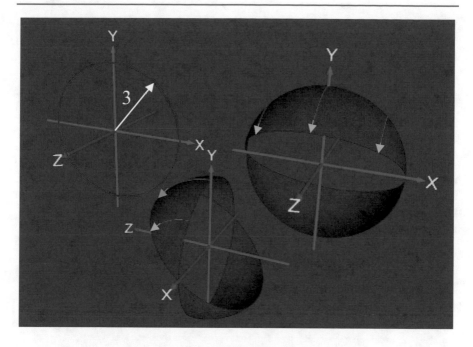

Fig. 2.30 Parametric definition of the sphere with radius 2

The upper part of the circle creates the front part of the sphere, while the lower part—the rear half—the rear half of the sphere. Let's define it mathematically using two parameters: u to define a circle and v to define the circle rotation. The circle is defined by

$$
\begin{aligned}
x &= 3.0\,cos(2\pi u) \\
y &= 3.0\,sin(2\pi u) \\
z &= 0 \\
u &\in [0,1]
\end{aligned}
\tag{2.63}
$$

Each point of this circle then rotates about axis X creating a half-circle, which can be defined by

$$
\begin{aligned}
y &= r\cos(\pi v) \\
z &= r\sin(\pi v) \\
v &\in [0,1]
\end{aligned}
\tag{2.64}
$$

where r is in fact the $y-$ coordinate, as it is defined in Eq. (2.63). We then will put together Eq. (2.63) and Eq. (2.64) by replacing r in Eq. (2.64) with its formula from Eq. (2.63) and we will also add the formula for the $x-$ coordinate:

$$
\begin{aligned}
x &= 3.0 \ cos(2\pi u) \\
y &= 3.0 \ sin(2\pi u) \ cos(\pi v) \\
z &= 3.0 \ sin(2\pi u) \ sin(\pi v) \\
u, v &\in [0,1]
\end{aligned}
\tag{2.65}
$$

This approach can be then used for making any other surface created by the rotation of any curve defined in the XY plane. Thus, we can define the cone displayed in Fig. 2.31 by rotation of a straight line segment about axis X by 2π angle.

If the segment is defined by

$$
\begin{aligned}
x &= 1.8u \\
y &= 2u \\
u &\in [0,1]
\end{aligned}
\tag{2.66}
$$

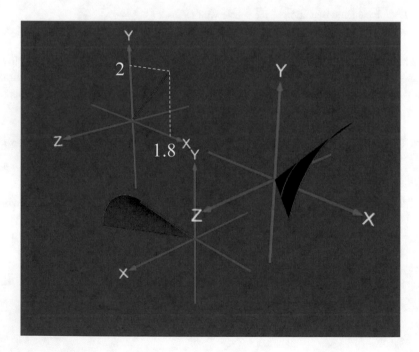

Fig. 2.31 Parametric definition of the sphere with radius 2

the parametric equations of the cone will be obtained by merely replacing $x-$ and $y-$ coordinate formulas in Eq. (2.65) with the $x-$ and $y-$ coordinate formulas from Eq. (2.66):

$$
\begin{aligned}
x &= 1.8u \\
z &= 2u\,\sin(\pi v) \\
z &= 2u\,\sin(\pi v) \\
u, v &\in [0,1]
\end{aligned}
\tag{2.67}
$$

This method of creating parametric definitions of surfaces is called *sweeping*. In fact, we previously already considered curves as created by moving points, or as points sweeping the curves. The surfaces can also be understood as being swept by curves which, in turn, may change their geometry while moving in 3D space.

This approach of defining surfaces by translation and rotation of the plain curves can be generalized as displayed in Fig. 2.32 and Fig. 2.33.

The mnemonic rule for making a surface by translation of a curve is to simply add the parametric definitions of a straight line segment to the definitions of the curve, however, it has to be remembered that the curve will be translated without changing its orientation, i.e. it will be a parallel translation along the straight line.

The mnemonic rule for making a surface by rotation of a plane curve about a coordinate axis is to /1/ copy to the final set of equations the formula defining the same coordinate of the curve as the axis of rotation /2/ write parametric definitions of a circle in the coordinate plane orthogonal to the axis of rotation, and /3/ replace the radius of rotation with the formula of the second coordinate of the curve.

Let's consider a few examples of making surfaces by the translational and rotational sweeping of curves.

A "rose" curve displayed in Fig. 2.34a is defined by an equation in polar coordinates by $r = 0.9\cos(5\alpha)$, $\alpha \in [0, 2\pi]$.

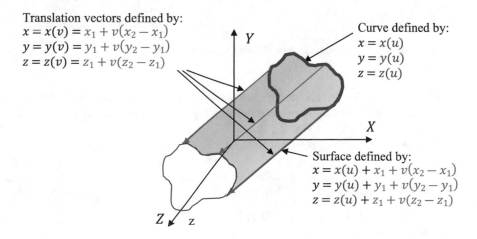

Translation vectors defined by:
$$
\begin{aligned}
x &= x(v) = x_1 + v(x_2 - x_1) \\
y &= y(v) = y_1 + v(y_2 - y_1) \\
z &= z(v) = z_1 + v(z_2 - z_1)
\end{aligned}
$$

Curve defined by:
$$
\begin{aligned}
x &= x(u) \\
y &= y(u) \\
z &= z(u)
\end{aligned}
$$

Surface defined by:
$$
\begin{aligned}
x &= x(u) + x_1 + v(x_2 - x_1) \\
y &= y(u) + y_1 + v(y_2 - y_1) \\
z &= z(u) + z_1 + v(z_2 - z_1)
\end{aligned}
$$

Fig. 2.32 Generalization of surface definition by translational sweeping of a curve

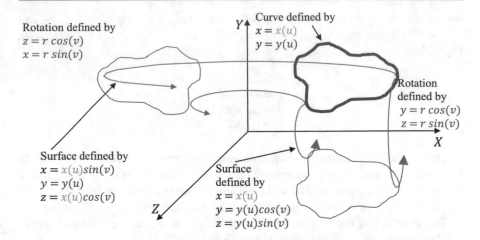

Rotation defined by
$z = r\,cos(v)$
$x = r\,sin(v)$

Curve defined by
$x = x(u)$
$y = y(u)$

Rotation
defined by
$y = r\,cos(v)$
$z = r\,sin(v)$

X

Surface defined by
$x = x(u)sin(v)$
$y = y(u)$
$z = x(u)cos(v)$

Surface
defined by
$x = x(u)$
$y = y(u)cos(v)$
$z = y(u)sin(v)$

Z

Fig. 2.33 Generalization of surface definition by rotational sweeping of a curve

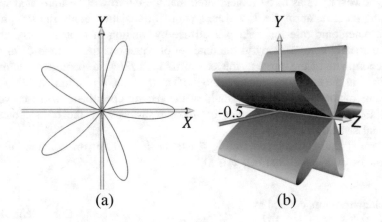

(a)   (b)

Fig. 2.34 A "rose" curve and a surface obtained by translating this curve

We have to define by parametric functions $x(u,v), y(u,v), u, v \in [0,1]$ the surface, which is obtained by translational sweeping of the curve in the direction of axis Z as it is displayed in Fig. 2.34b.

From the definitions of the right triangle:$x = r\cos\alpha, y = r\sin\alpha$

$$x = 0.9\cos(5\alpha)\cos\alpha$$
$$y = 0.9\cos(5\alpha)\sin\alpha \qquad\qquad (2.68)$$
$$\alpha \in [0, 2\pi]$$

$$x = 0.9\cos(5u2\pi)\cos(u2\pi) = 0.9\cos(u10\pi)\cos(u2\pi)$$
$$y = 0.9\cos(5u2\pi)\sin(u2\pi) = 0.9\cos(u10\pi)\cos(u2\pi) \qquad (2.69)$$
$$u \in [0, 1]$$

The surface then is obtained by adding the following function for the z-coordinate:

$$z = -0.5 + v(1-(-0.5)) = -0.5 + 1.5v \quad v \in [0, 1] \qquad (2.70)$$

In the next example, the sine curve displayed in Fig. 2.35a is rotated about axis Y by 1.5π to make a surface displayed in Fig. 2.35b. To define this surface parametrically, we will first write parametric equations $x(u), y(u), u[0, 1]$ defining the curve in the coordinate plane XY:

$$x = 0.7u + 0.3$$
$$y = 0.25\sin(4\pi u) + 0.25 \qquad (2.71)$$
$$z = 0.0$$

Next, we will obtain parametric equations $x(u, v), y(u, v), z(u, v), u, v, \in [0, 1]$ defining the surface:

$$x = (0.7u + 0.3)\sin(v \cdot 2 \cdot 0.75\pi + \pi/2)$$
$$y = 0.25\sin(4\pi u) + 0.25 \qquad (2.72)$$
$$z = (0.7u + 0.3)\cos(v \cdot 2 \cdot 0.75\pi + \pi/2)$$

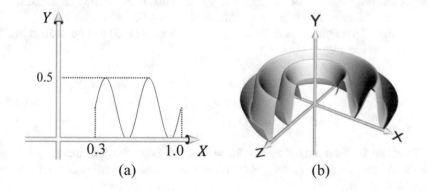

Fig. 2.35 Making a surface by rotating a sine curve

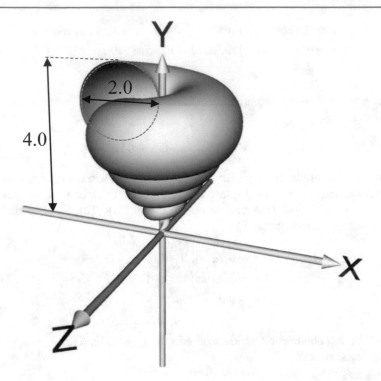

Fig. 2.36 Making a surface by rotation and translation of a circle

Let's consider one more complex example of making a surface of a spirally coiled snail displayed in Fig. 2.36. The surface is created by both rotational and translational sweeping of a circle. The circle makes 5 full spiral revolutions while always touching the axis of rotation. The circle radius is expanding from 0.0 to 1.0 at the final location. The total height of the shell is 4.0.

A circle with a radius 1 and the center at coordinates $(1, 0)$ can be defined in XY coordinate plane by

$$
\begin{aligned}
x &= 1 \cdot \cos(2\pi u) + 1 \\
y &= 1 \cdot \sin(2\pi u) \\
u &\in [0, 1]
\end{aligned}
\tag{2.73}
$$

To rotate it about axis Y by $5 \times 2\pi = 10\pi$, we copy the formula for $y-$ coordinate and use the formula for $x-$ coordinate as a variable radius of rotation. The rotation has to be defined in the right-hand ZX system. Since it will start on axis Z, we have to use a negative offset to start the rotation on plane XY:

$$
\begin{aligned}
x &= (1\ cos(2\pi u) + 1)\ sin(10\pi v - \pi/2) \\
y &= 1\ sin(2\pi u) \\
z &= (1\ cos(2\pi u) + 1)\ cos(10\pi v - \pi/2) \\
u, v &\in [0,1]
\end{aligned}
\tag{2.74}
$$

To make the circle change its radius from 0 to 1 while always touching the axis of rotation, we will use parameter v which controls the rotation in place of the constant radius 1 and offset 1:

$$
\begin{aligned}
x &= (v\ cos(2\pi u) + v)\ sin(10\pi v - \pi/2) \\
y &= v\ sin(2\pi u) \\
z &= (v\ cos(2\pi u) + v)\ cos(10\pi v - \pi/2) \\
u, v &\in [0,1]
\end{aligned}
\tag{2.75}
$$

Finally, to translate the circle along axis Y we will make a variable offset changing from 0 to 3 also controlled by parameter v:

$$
\begin{aligned}
x &= (v\ cos(2\pi u) + v)\ sin(10\pi v - \pi/2) \\
y &= v\ sin(2\pi u) + 3v \\
z &= (v\ cos(2\pi u) + v)\ cos(10\pi v - \pi/2) \\
u, v &\in [0,1]
\end{aligned}
\tag{2.76}
$$

We will revisit the topic of sweeping in Chapter 3 and will see how any curve, not only plane curves, can be used for sweeping surfaces using matrix transformations.

2.5 Solid Objects

When sampling curves, points have only one DOF to move—forward and backward. When sampling surfaces, points can in addition move sideways which adds one more DOF. When sampling solid objects, points can further move inside and outside their surfaces, and this is the third DOF. We will, therefore, think of solid objects as set of points created by all their possible locations on the surfaces of objects as well as inside or on one side of them. We may also represent solid objects as sets of points created by moving surfaces. Solid objects can be then visualized either by sampling and displaying individual points that form the object, or by displaying polygons (usually, triangles) that interpolate surfaces bounding the solid object.

Representing solid objects by points is done in a form of *voxels*. Voxels are volumetric pixels or Volume Picture Elements that usually represent a value on a regular grid in 3D space. Each voxel can be displayed either as an individual pixel or as a group of them. Different colors of voxels may visualize actual visible colors of objects as well as their physical properties such as density.

Since storing large voxel data sets is expensive in terms of memory usage, the same linear interpolation may become very handy here. By expanding the principles of bilinear interpolation to 3D solids, and by introducing the third parameter, so-called *Trilinear Interpolation* can be defined. It can be used for calculating the values of voxel colors between the grid cells (Fig. 2.37):

$$\mathbf{P}' = \mathbf{P}_1 + u(\mathbf{P}_2 - \mathbf{P}_1); \quad \mathbf{P}'' = \mathbf{P}_3 + u(\mathbf{P}_4 - \mathbf{P}_3); \quad \mathbf{P} = \mathbf{P}' + v(\mathbf{P}'' - \mathbf{P}')$$
$$\mathbf{R}' = \mathbf{R}_1 + u(\mathbf{R}_2 - \mathbf{R}_1); \quad \mathbf{R}'' = \mathbf{R}_3 + u(\mathbf{R}_4 - \mathbf{R}_3); \quad \mathbf{R} = \mathbf{R}' + v(\mathbf{R}'' - \mathbf{R}')$$
$$\mathbf{V} = \mathbf{P} + w(\mathbf{R} - \mathbf{P})$$

$$(2.77)$$

When displaying solid objects by polygons, we will always see the object boundaries wherever we cut the object, while when displaying surfaces we can see their inner and outer parts, as illustrated in Fig. 2.38 where a solid hemisphere and a surface hemisphere are displayed.

2.5.1 Defining Solids by Parametric Functions

To define algorithmically the locations of points on and inside the solid objects, we use the definition by mathematical functions. Parametric definitions of curves and surfaces can easily be converted to definitions of solid objects if we add an additional parameter to control one more degree-of-freedom. Let's consider a few examples.

We will start with plain objects. To define plain areas bounded by curves, we have to add one more parameter to the curve definitions. Thus, a circle turns into a plane *circular disk* (Fig. 2.39a) if we make a variable radius:

$$\begin{aligned} x &= vR \cos 2\pi u \\ y &= vR \sin 2\pi u \\ u, v &\in [0,1] \end{aligned}$$

$$(2.78)$$

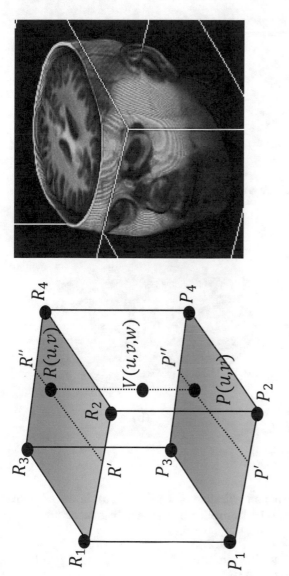

Fig. 2.37 Trilinear interpolation of voxels and voxel image

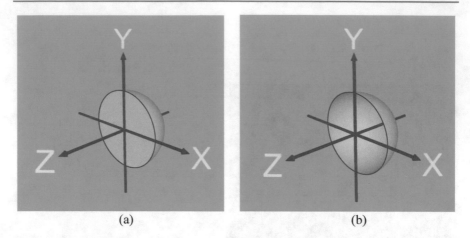

Fig. 2.38 Displaying (a) solid hemisphere and (b) surface hemisphere

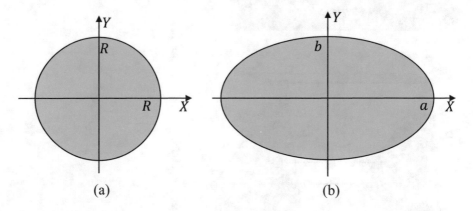

Fig. 2.39 a Circular disk and **b** elliptical disk

In the same way, we can define an *elliptical disk* (Fig. 2.39b), however, both variable semi-axes a and b have to be controlled by the same parameter v:

$$
\begin{aligned}
x &= va \cos 2\pi u \\
y &= vb \sin 2\pi u \\
u, v &\in [0,1]
\end{aligned}
\tag{2.79}
$$

Both plane disks can then be converted into 3D solid cylinders if we add the third coordinate and displace them as we did with the surfaces. Thus, the solid

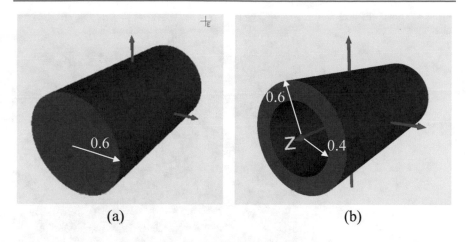

Fig. 2.40 **a** Solid cylinder and **b** solid cylinder with a hole

circular cylinder with radius 0.6 aligned with axis Z so that its $z-$ coordinate is spanning from -0.5 to 1.0 (Fig. 2.40a) can be defined by

$$
\begin{aligned}
x &= 0.6v \cos 2\pi u \\
y &= 0.6v \sin 2\pi u \\
z &= -0.5 + 1.5w \\
u, v, w &\in [0,1]
\end{aligned}
\tag{2.80}
$$

We can also make a cylindrical hole in it (Fig. 2.40b) by making the radius values linearly change from 0.4 to 0.6:

$$
\begin{aligned}
x &= (0.4 + 0.2v) \cos 2\pi u \\
y &= (0.4 + 0.2v) \sin 2\pi u \\
z &= -0.5 + 1.5w \\
u, v, w &\in [0,1]
\end{aligned}
\tag{2.81}
$$

A solid elliptical torus (Fig. 2.41) can be defined by rotation about axis Y the plain elliptical disk with semi-axes 0.5 and 1.0 centered at $(3, 0)$:

$$
\begin{aligned}
\text{Disk}: x &= 0.5v \cos(2\pi u) + 3.0 \\
y &= 1.0v \sin(2\pi u)
\end{aligned}
\tag{2.82}
$$

Fig. 2.41 Solid elliptical torus

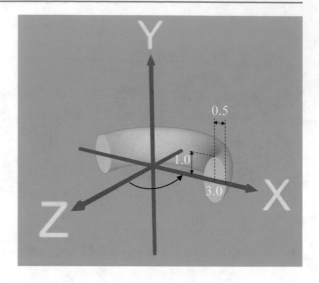

$$
\begin{aligned}
\text{Torus}: x &= (0.5v\cos(2\pi u) + 3.0)\sin(w\pi + \pi/2) \\
y &= 1.0v\sin(2\pi u) \\
z &= (0.5v\cos(2\pi u) + 3.0)\cos(w\pi + \pi/2) \\
u, v, w &\in [0, 1]
\end{aligned} \tag{2.83}
$$

In the same way, we can define a solid sphere and a solid ellipsoid as well as other previously considered surfaces to solid objects. Thus, the solid sphere with radius 3.0 can be defined by modifying (2.65) to make the radius controlled by the third parameter w:

$$
\begin{aligned}
x &= 3.0w\,cos(2\pi u) \\
y &= 3.0w\,sin(2\pi u)\,cos(\pi v) \\
z &= 3.0w\,sin(2\pi u)\,sin(\pi v) \\
u, v, w &\in [0,1]
\end{aligned} \tag{2.84}
$$

The equations of the solid ellipsoid with the semi-axes 1.0, 0.8, and 0.5 can be defined by

$$
\begin{aligned}
x &= 1.0w\,cos(2\pi u) \\
y &= 0.8w\,sin(2\pi u)\,cos(\pi v) \\
z &= 0.5w\,sin(2\pi u)\,sin(\pi v) \\
u, v, w &\in [0,1]
\end{aligned} \tag{2.85}
$$

Fig. 2.42 Solid cone

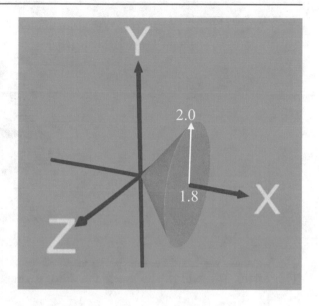

A conical surface defined by (2.67) after a little modification becomes a solid cone (Fig. 2.42):

$$
\begin{aligned}
x &= 1.8u \\
y &= 2uw\ cos(2\pi v) \\
z &= 2uw\ sin(2\pi v) \\
u, v, w &\in [0,1]
\end{aligned}
\tag{2.86}
$$

Let's consider two more examples of creating solid objects based on only a few formulas that we learnt. In the first example, we will make a solid pyramid displayed in Fig. 2.43.

The steps of making it are displayed in Fig. 2.44.

First, we make the square polygon which will become the base of the pyramid (Fig. 2.44a). This requires two parameters to be used:

$$
\begin{aligned}
x &= -1 + 2u \\
y &= 0 \\
z &= -1 + 2v \\
u, v &= [0,1]
\end{aligned}
\tag{2.87}
$$

Fig. 2.43 Solid pyramid

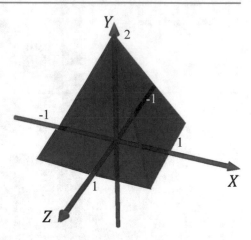

After that, we translate this polygon by 2 units up to make a solid cube (Fig. 2.44b). This will require to add the third parameter w:

$$\begin{aligned}
x &= -1 + 2u \\
y &= 2w \\
z &= -1 + 2v \\
u, v, w &= [0,1]
\end{aligned}$$

(2.88)

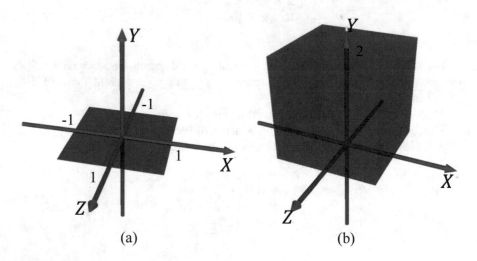

(a)   (b)

Fig. 2.44 Steps of making the solid pyramid

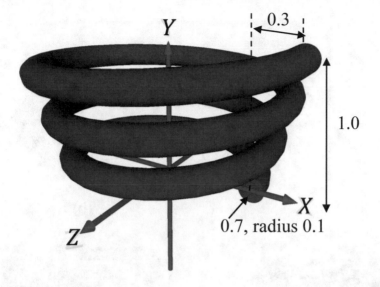

Fig. 2.45 Solid helix

Finally, the cube is converted to the pyramid by linear reduction of the base while moving up:

$$
\begin{aligned}
x &= (-1 + 2u)(1 - w) \\
y &= 2w \\
z &= (-1 + 2v)(1 - w) \\
u, v, w &= [0,1]
\end{aligned} \tag{2.89}
$$

In the next example, we will make a solid helix displayed in Fig. 2.45 while the steps of making it are displayed in Fig. 2.46.

First, in Fig. 2.46a we define a solid disk with the center at point (0.7, 0, 0) and radius 0.1:

$$
\begin{aligned}
x &= 0.1v \, cos(2\pi u) + 0.7 \\
y &= 0.1v \, sin(2\pi u) \\
z &= 0 \\
u, v &= [0,1]
\end{aligned} \tag{2.90}
$$

Next, in Fig. 2.46b we rotate this disk about axis Y to make a solid torus which requires to add one more parameter w:

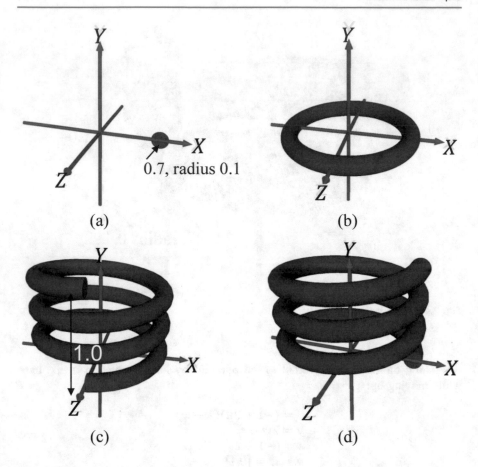

Fig. 2.46 The steps of making the solid helix

$$x = (0.1v\,cos(2\pi u) + 0.7)\,sin(2\pi w)$$
$$y = 0.1v\,sin(2\pi u)$$
$$z = (0.1v\,cos(2\pi u) + 0.7)\,cos(2\pi w)$$
$$u, v, w = [0,1]$$

(2.91)

Three revolutions together with vertical translation are made in Fig. 2.46c:

$$x = (0.1v\,cos(2\pi u) + 0.7)\,sin(6\pi w)$$
$$y = 0.1v\,sin(2\pi u) + 1w$$
$$z = (0.1v\,cos(2\pi u) + 0.7)\,cos(6\pi w)$$
$$u, v, w = [0,1]$$

(2.92)

In Fig. 2.46d, the whole shape is rotated about axis Y by $\frac{\pi}{2}$

$$
\begin{aligned}
x &= (0.1v\,cos(2\pi u) + 0.7)\,sin(6\pi w + \frac{\pi}{2}) \\
y &= 0.1v\,sin(2\pi u) + 1w \\
z &= (0.1v\,cos(2\pi u) + 0.7)\,cos(6\pi w + \frac{\pi}{2}) \\
u,v,w &= [0,1]
\end{aligned}
\tag{2.93}
$$

Finally, the rotation radius of the helix is made variable:

$$
\begin{aligned}
x &= (0.1v\,cos(2\pi u) + 0.7 + 0.3w)\,sin(6\pi w + \frac{\pi}{2}) \\
y &= 0.1v\,sin(2\pi u) + 1w \\
z &= (0.1v\,cos(2\pi u) + 0.7 + 0.3w)\,cos(6\pi w + \frac{\pi}{2}) \\
u,v,w &= [0,1]
\end{aligned}
\tag{2.94}
$$

2.5.2 Constructive Solid Geometry by Functions

Implicit functions check whether the equality $f(x,y,z) = 0$ holds for any sampled point. Those points whose coordinates satisfy the equation belong to the surface of the shape, while all other are considered to be located outside it. If we change the implicit function to a predicate

$$
f(x,y,z) \geq 0 \ \text{ or } \ f(x,y,z) \leq 0
\tag{2.95}
$$

we will then mathematically define a set of points that are not only on the surface but also on one side of the surface. For closed surfaces, we will define points inside or outside the surface, while for surfaces which are not closed we will define infinite half-spaces on one of the surface sides. If we are not only interested in examining the sign of the functions Eq. (2.95) but rather in its value, it becomes an explicit function of Cartesian coordinates in two- or three-dimensional space. This value can give us, for example, a distance to the surface or some physical property (e.g., density) of the modeled material:

$$
\begin{aligned}
g = f(x,y) \geq 0 \ \text{ or } \ g = f(x,y) \leq 0 \\
g = f(x,y,z) \geq 0 \ \text{ or } \ g = f(x,y,z) \leq 0
\end{aligned}
\tag{2.96}
$$

We will consider only definitions

$$
g = f(x,y) \geq 0 \ \text{ and } \ g = f(x,y,z) \geq 0
\tag{2.97}
$$

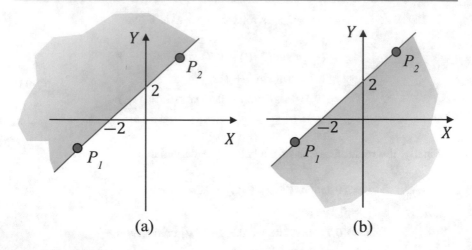

Fig. 2.47 Examples of definitions of half-plane bounded by line

which state that points with coordinates (x, y) or (x, y, x) belong to the surface of the shape if the value of the function Eq. (2.97) is equal to 0; located inside the shape (or on a specific side of its boundary) if the value of the function is greater than 0, and do not belong to the shape if the value of the function is less than zero. The scientific name of such functions is *FRep* (*Function Representation*) [1], and all further math will work only for these functions where "greater than equal" predicate is used.

Let's examine what happens when implicit functions are converted to *FReps*. An implicit equation of a line becomes a function defining half-plane bounded by this line. The upper (Fig. 2.47a) or lower (Fig. 2.47b) half of the space above or below the line can then be easily toggled by changing the function sign:

$$\frac{x}{-2} + \frac{y}{2} = 1 \Rightarrow \frac{x}{-2} + \frac{y}{2} - 1 = 0 \Rightarrow \frac{x}{-2} + \frac{y}{2} - 1 \geq 0$$
$$-\left(\frac{x}{-2} + \frac{y}{2} - 1\right) = 0 = \frac{x}{2} - \frac{y}{2} + 1 \geq 0 \tag{2.98}$$

In the same way, we can define a half-space on one or another side of a plane by writing its equation in intercepts (Fig. 2.48a) and then inverting it by changing the sign in front of the function to negative (Fig. 2.48b):

$$\frac{x}{2} + \frac{y}{2} + \frac{z}{2} = 1 \Rightarrow \frac{x}{2} + \frac{y}{2} + \frac{z}{2} - 1 \geq 0$$
$$-\left(\frac{x}{2} + \frac{y}{2} + \frac{z}{2} - 1\right) = -\frac{x}{2} - \frac{y}{2} - \frac{z}{2} + 1 \geq 0 \tag{2.99}$$

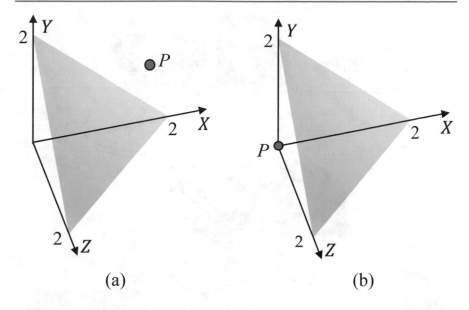

Fig. 2.48 Examples of definitions of half-space bounded by plane

Implicit functions defining a sphere with radius r and an ellipsoid with semi-axes a, b, c can be then modified to *FRep* functions defining a solid sphere and a solid ellipsoid:

$$r^2 - x^2 - y^2 - z^2 \geq 0$$
$$1 - \left(\tfrac{x}{a}\right)^2 - \left(\tfrac{y}{b}\right)^2 - \left(\tfrac{z}{c}\right)^2 \geq 0 \tag{2.100}$$

In the same way, we may continue with any implicit function only paying attention to the side of the shape boundary which will be defined by a simple change of the equality to the inequality. If another side is required to be defined, we have to multiply the whole defining function by -1.

An ability to define individual solid objects by *FRep* functions can be greatly extended if we consider what is called in shape modeling "*Constructive Solid Geometry*" or *CSG*. CSG is a family of schemes representing solid objects as Boolean constructions and combinations of solid components using three basic operations *union* $\cup$, *intersection* $\cap$, and *difference* \. In CSG, the construction of objects can be visually represented as a binary tree, called *CSG tree*. Each leaf in this tree is a simple (primitive) object, and each non-terminal node is either a *Boolean operation* or some transformation that operates on the sub-nodes. An example of a CSG tree is shown in Fig. 2.49 where a solid L-shaped block with a cylindrical hole in it is built by unifying two cuboids followed by subtraction of a cylinder from the resulting shape.

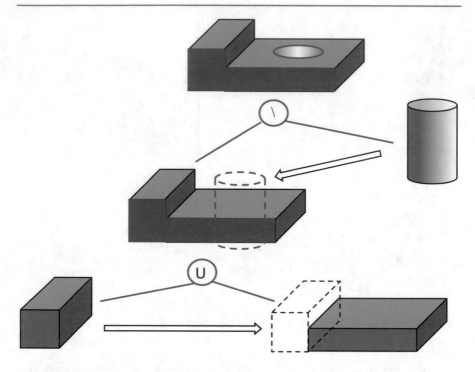

Fig. 2.49 CSG tree

There are different ways how to implement such CSG trees with a computer. In this book, we define everything with simple formulas. The Boolean operations (also called *Set-Theoretic Operations* or *STO*) then will become mathematical operators transforming *FRep* functions defining primitive objects (inputs) to the function defining their Boolean combination. One of the ways how it can be done is as follows Complement::

$$
\begin{array}{lll}
\text{Input}: & G_1 : f_1(x,y,z) \geq 0, & G_2 : f_2(x,y,z) \geq 0 \\
\text{Union}: & G = G_1 \cup G_2 : f(x,y,z) = \max(f_1,f_2) \geq 0 \\
\text{Intersection}: & G = G_1 \cap G_2 : f(x,y,z) = \min(f_1,f_2) \geq 0 & \quad (2.101) \\
\text{Complement}: & G = \overline{G_1} \quad f(x,y,z) = -f_1 \geq 0 \\
\text{Difference}: & G = G_1 \backslash G_2 : f(x,y,z) = \min(f_1,-f_2) \geq 0
\end{array}
$$

For example, a complex solid shape G_5 is constructed by the intersection of shapes G_1 with G_2 followed by the union of the result with shape G_3 followed subtraction of shape G_4 from the resulting shape:

$$
\begin{aligned}
G_5 &= (G_3 \cup (G_1 \cap G_2)) \backslash G_4 \\
f_5(x,y,z) &= \min(\max(f_3, \min(f_1,f_2)), -f_4) \geq 0
\end{aligned}
\qquad (2.102)
$$

Let's consider a few examples. In Fig. 2.50 we construct several two-dimensional shapes. In Fig. 2.50a, a circular disk with radius 1 is subtracted from the elliptical disk with semi-axes 3 and 2. In Fig. 2.50b, the elliptical disk is replaced with a rectangular polygon which is created by the intersection of four half-planes $x \geq -3$, $x \leq 3$, $y \geq -2$, and $y \leq 2$. These inequalities are then converted to functions $f(x, y) \geq 0$:

$$x + 3 \geq 0; \quad 3 - x \geq 0; \quad y + 2 \geq 0; \quad 2 - y \geq 0 \qquad (2.103)$$

Furthermore, in Fig. 2.50c, the circular disk is translated by 1 in positive x direction. In Fig. 2.50d, the circular disk is further translated by 2 more units in positive x direction with the accumulated translation offset 3, and the subtraction

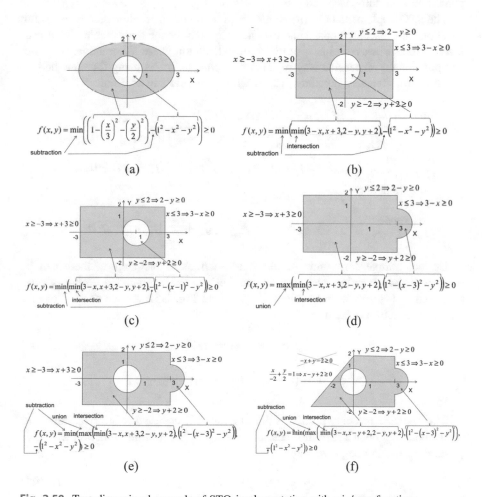

Fig. 2.50 Two-dimensional example of STO implementation with min/max functions

operation is replaced with the union operation (*max* function). Finally in Fig. 2.50e, the half-plane $x \geq -3$ is replaced with the half-plane bounded by the line passing through the points with coordinates (-2,0) and (0,2). Out of two possible equations of this line, we convert to *FRep* function the one which will define positive points below the line:

$$x - y + 2 \geq 0 \tag{2.104}$$

We then continue our exercise in three-dimensional space. In Fig. 2.51a, we define a solid cube by intersecting 6 plane-bounded half-spaces. In Fig. 2.51b, we add to the cube a sphere with a radius 0.5 and the center at coordinates (1, 1, 1). Finally, in Fig. 2.51c, we subtract from the shape a cylinder which is obtained by the intersection of the cylinder with a radius 0.5 aligned with axis Z and a plane-bounded half-space and $z > 0$.

The STO implementation with min/max function has a problem that the resulting functions are discontinuous. This discontinuity may create undesirable visual artifacts on the surfaces of objects at the color calculation phase. Alternative, but more computationally expensive way of STO implementation, is using the so-called *R*-functions (or Rvachev functions) [2, 3] which provide any continuity C^m:

$$
\begin{array}{ll}
\text{Input } G_1, G_2: & f_1(x,y,z) \geq 0, \ f_2(x,y,z) \geq 0 \\
\text{Union :} & \\
G = G_1 \cup G_2: & f(x,y,z) = \left(f_1 + f_2 + \sqrt{f_1^2 + f_2^2} \right) \left(f_1^2 + f_2^2 \right)^{\frac{m}{2}} \geq 0 \\
\text{Intersection :} & \\
G = G_1 \cap G_2: & f(x,y,z) = \left(f_1 + f_2 - \sqrt{f_1^2 + f_2^2} \right) \left(f_1^2 + f_2^2 \right)^{\frac{m}{2}} \geq 0 \\
\text{Complement :} & \\
G = \overline{G_1}: & f(x,y,z) = -f_1 \geq 0 \\
\text{Difference :} & \\
G = G_1 \backslash G_2: & f(x,y,z) = \left(f_1 - f_2 - \sqrt{f_1^2 + f_2^2} \right) \left(f_1^2 + f_2^2 \right)^{\frac{m}{2}} \geq 0
\end{array}
\tag{2.105}
$$

Besides simple conversion of implicit functions to inequalities, there can be other ways of obtaining such definitions. Let's consider so-called *blobby shapes* proposed in [4] and further evolved in [5–7]. In Fig. 2.52 we can see a plot of a Gaussian function defined by

$$g = f(x,y,z) = ae^{-r} \tag{2.106}$$

$$\text{where} \quad r = (x - x_b)^2 + (y - y_b)^2 + (z - z_b)^2$$

This function has only positive values however if we use any positive value c as a threshold and deduct it from function Eq. (2.106) the resulting function will have positive, zero, and negative values.

$$g = f(x,y,z) = ae^{-r} - c \tag{2.107}$$

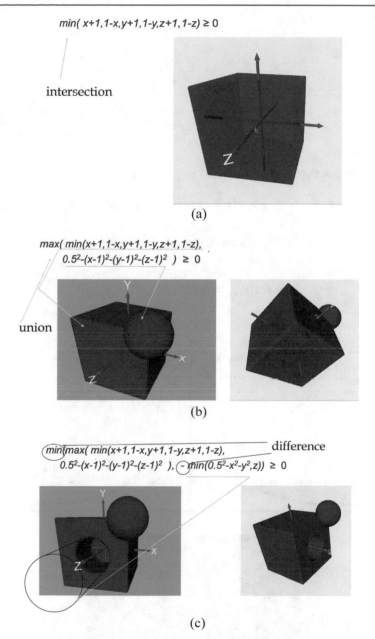

Fig. 2.51 Three-dimensional example of STO implementation with min/max functions

Fig. 2.52   Gaussian function

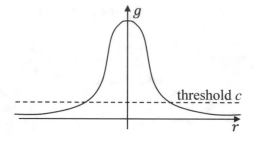

This function then can be converted to *FRep* form:

$$g = f(x,y,z) = ae^{-r} - c \geq 0 \qquad (2.108)$$

and when it is visualized it will be displayed as a sphere (Fig. 2.53):

If we sum up two such functions

$$g = f(x,y,z) = a_1 e^{-r_1} + a_2 e^{-r_2} - c \geq 0 \qquad (2.109)$$

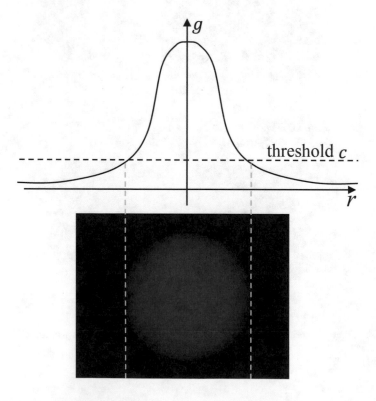

Fig. 2.53   Blobby shape function and its visualization

then the resulting shape will look as it is displayed in Fig. 2.54a. Changing values r_1, r_2, which define placements of the shapes, and parameters a_1, a_2 which define size of the shapes, will make different appearances of the resulting shape as illustrated in Fig. 2.54b. The individual shapes, defined by Eq. (2.106), appear to act like liquid droplets. When they are positioned close to each other, they stick together and form a bigger droplet. This is why the scientific name given to such shapes is *blobby shapes* or *blobs* or *soft objects*. They can be very efficiently used for making shapes with smooth boundaries between the individual blobs, as

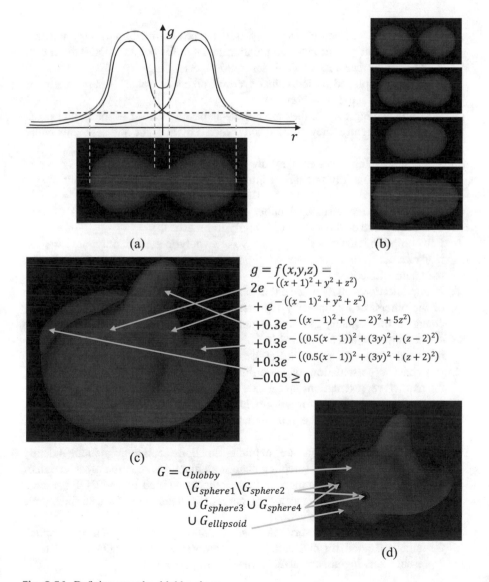

(a)   (b)

$$g = f(x,y,z) =$$
$$2e^{-((x+1)^2 + y^2 + z^2)}$$
$$+ e^{-((x-1)^2 + y^2 + z^2)}$$
$$+0.3e^{-((x-1)^2 + (y-2)^2 + 5z^2)}$$
$$+0.3e^{-((0.5(x-1))^2 + (3y)^2 + (z-2)^2)}$$
$$+0.3e^{-((0.5(x-1))^2 + (3y)^2 + (z+2)^2)}$$
$$-0.05 \geq 0$$

(c)

$$G = G_{blobby}$$
$$\backslash G_{sphere1} \backslash G_{sphere2}$$
$$\cup G_{sphere3} \cup G_{sphere4}$$
$$\cup G_{ellipsoid}$$

(d)

Fig. 2.54 Defining complex blobby shape

illustrated in Fig. 2.54c. These shapes, since they are defined by functions $g = f(x, y, z) \geq 0$, can then be used together with other *FRep* shapes which we considered in this chapter, as it is illustrated in Fig. 2.54d where other shapes are added to and subtracted from the blobby shape.

2.6 Summary

1. We classified geometric shapes by the degrees-of-freedom (DOF) which a geometric point may have to sample them: 0 DOF for points, 1 DOF for curves, 2 DOF for surfaces, and 3 DOF for solid objects.
2. Points are displayed as individual colored pixels or splats (little triangles or disks always facing the observer).
3. When displaying curves, they are either displayed as a set of individual pixels or, more frequently, they are first interpolated by straight line segments called polylines.
4. When displaying surfaces, they are either displayed by computing colors of individual pixels (ray tracing) or interpolated by triangular polygons (polygon mesh).
5. Solid objects are visualized either as a set of voxels (3D pixels) or their boundary surfaces are displayed as polygon meshes.
6. For computing locations of pixels, voxels, polylines, and polygons, we use mathematical representations of geometric shapes by implicit, explicit, and parametric functions.
7. In implicit representation, we either cannot express one coordinate as a function of the others, or intentionally represent the function in the form $f(\mathbf{P}) = 0$. By changing this equality into an inequality $g = f(\mathbf{P}) \geq 0$, we define not only a surface but the space bounded by this surface, or a half-space. Implicit functions can efficiently represent Set-theoretic (Boolean) operations.
8. In explicit representation, one coordinate is a function of the others. Usually, the explicit representations create axis dependency, i.e., there may be no single representation available for any orientation of the shape or multivalued functions will be used in the representation which is not a desirable case for visualization algorithms.
9. Parametric representation are explicit functions computing independently coordinates of the points sampling shapes as functions of some other variables called parameters. The number of parameters is defined by the DOF for sampling the shapes: one parameter for a curve, two parameters for a surface, three parameters for a solid object.
10. Parametric formulas are very efficient for drawing while explicit and implicit formulas are good for calculating shape properties. We usually use all three representations together to achieve maximum performance.

References

1. Pasko A, Adzhiev V, Sourin A, Savchenko V. Function representation in geometric modeling: concepts, implementation and applications. The Visual Computer, Springer, 11(8): 429–446, 1995
2. V.L. Rvachev, "On the analytical description of some geometric objects", *Reports of Ukrainian Academy of Sciences*, vol. **153**, no. 4, 1963, pp. 765–767 (in Russian).
3. V. Shapiro, Semi-analytic geometry with R-Functions, Acta Numerica, Cambridge University Press, 2007, 16: 239–303
4. J. Blinn. *A Generalization of Algebraic Surface Drawing.*, ACM Transactions on Graphics. 1982.
5. G. Wyvill, C. McPheeters, B. Wyvill. *Data Structure for soft objects.* Visual Computer. 1986
6. G. Wyvill, C. McPheeters, B. Wyvill. *Animating soft objects.* Visual Computer. 1986
7. S. Muraki. *Volumetric Shape Description of Range Data using "Blobby Model".* ACM SIGGRAPH. 1991.

Transformations

3

3.1 Mathematics of Transformations

We sample shapes as individual points defined by their coordinates. If we need to transform shapes, then all the point coordinates have to be changed according to some transformation functions:

$$\mathbf{P}' - f(\mathbf{P}) \tag{3.1}$$

where $\mathbf{P}$ is either a *row position vector* of points defined by their coordinates

$$\mathbf{P} = [x \quad y] \text{ or } \mathbf{P} = [x \quad y \quad z] \tag{3.2}$$

or a *column position vector*

$$\mathbf{P} = \begin{bmatrix} x \\ y \end{bmatrix} \text{ or } \mathbf{P} = \begin{bmatrix} x \\ y \\ z \end{bmatrix} \tag{3.3}$$

Depending on the type of transformation, the defining functions interpolating the shape (which we learnt in Chap. 2) may then either remain basically the same with very little corrections or have to be completely changed. Transformations like translation, rotation and reflection do not change the shapes but only their location and orientation. These transformations are called *Euclidean transformations* [1, 2]. They preserve length and angles, and, therefore, lines will transform to lines, circles to circles, spheres to spheres, etc. However, when scaling is added to Euclidean transformations, while lines still will be transformed to lines, circles may change to ellipses, spheres to ellipsoids, etc. The shapes, however, still will look similar and recognizable. These transformations defined by combinations or translation, rotation, and scaling transformations are called *affine transformations* (from the Latin,

© The Author(s), under exclusive license to Springer Nature Switzerland AG 2021   85
A. Sourin, *Making Images with Mathematics*, Undergraduate Topics
in Computer Science, https://doi.org/10.1007/978-3-030-69835-5_3

affinis, "connected with"). They preserve parallel lines while not preserving sizes and angles. It still can be possible to use the same interpolating function with a proper selection of the reference points defining the shape. Alternatively, the defining function has to be applied first to generate the necessary number of points interpolating the shape, and these points then have to be transformed by the affine transformations. The number of points before and after Euclidean and affine transformations will remain the same. There can be also other transformations that are not defined by a combination of translation, rotation, and scaling transformations and which may even change the number of points after the transformations.

Let's consider affine transformations as a generalization of Euclidean transformations. Mathematically, 2D and 3D affine transformations can be defined as transformations of coordinates by linear functions—the very same functions that we used in Chap. 2 for defining straight lines, planes, and linear interpolation transformations of the shape parameters such as radii, lengths, etc.:

$$\begin{aligned} x' &= a\,x + b\,y + l \\ y' &= c\,x + d\,y + m \end{aligned} \tag{3.4}$$

$$\begin{aligned} x' &= a\,x + b\,y + c\,z + l \\ y' &= d\,x + e\,y + f\,z + m \\ z' &= g\,x + h\,y + k\,z + n \end{aligned} \tag{3.5}$$

To perform transformations, these functions can be written in the software code one by one for each of the transformations to be done, however, this approach may create a problem when the number of points to be transformed is large. Indeed, if the time needed to apply the transformation functions to the whole set of points is T, with n subsequent transformations it will become nT, i.e. it will grow proportionally to the number of subsequent transformations. To keep the transformation time the same regardless of the number of transformations used, matrix representation of transformations has to be used. It is known in mathematics and physics that several matrices can be combined into one matrix by multiplying them. Also, transformations based on matrix operations can be efficiently implemented with a computer, since for every point to be transformed, the same procedure will be called which will multiply the point coordinates by the transformation matrix. Therefore, if we find a way how to define each of the subsequent n transformations as a matrix $\mathbf{M}_i$, $i = 1, \ldots, n$, to apply these transformations to all the points $\mathbf{P}$ we have to first compute one transformation matrix $\mathbf{M} = \mathbf{M}_n \mathbf{M}_{n-1} \ldots \mathbf{M}_1$ and then apply it only one time to the points $\mathbf{P}$ as $\mathbf{P}' = \mathbf{MP}$. The time τ needed to compute the matrix product is significantly smaller than the time T needed to multiply the resulting matrix by many point coordinates ($\tau \ll T$). Therefore, the application of any number of transformations will require almost the same time provided the number of points to be transformed is very large (thousands of millions) which is a common case of computer graphics visualization.

3.2 Matrix Representation of Affine Transformations

Let's see how affine transformations can be defined using matrices. We will consider the representation of transformations with a column-represented position vector $\mathbf{P}$ as in Eq. (3.3). In that case, according to the rules of matrix multiplication, the matrix transformations in 2D and 3D spaces have to be written as

$$\mathbf{P}' = \begin{bmatrix} x' \\ y' \end{bmatrix} = \begin{bmatrix} a\,x + b\,y \\ c\,x + d\,y \end{bmatrix} = \begin{bmatrix} a & b \\ c & d \end{bmatrix} \begin{bmatrix} x \\ y \end{bmatrix} \tag{3.6}$$

$$\mathbf{P}' = \begin{bmatrix} x' \\ y' \\ z' \end{bmatrix} = \begin{bmatrix} a\,x + b\,y + c\,z \\ d\,x + e\,y + f\,z \\ g\,x + h\,y + k\,z \end{bmatrix} = \begin{bmatrix} a & b & c \\ d & e & f \\ g & h & k \end{bmatrix} \begin{bmatrix} x \\ y \\ z \end{bmatrix} \tag{3.7}$$

However, both the 2×2 and 3×3 matrices have no room for l, m, and n. They require additional columns to place these parameters, i.e. the matrices have to increase their size to 3×3 and 4×4, respectively (Fig. 3.1).

3.2.1 Homogeneous Coordinates

According to the matrix multiplication rules, an addition of one more column will require to add one more row to the position vector $\mathbf{P}$ for the coordinate which we do not have. Then, an elegant mathematical solution called *homogeneous coordinates* comes to help [3]. We will add one more artificial dummy coordinate w to the position vector in a way how it is defined in the following equations:

$$\mathbf{P}' = \begin{bmatrix} x' \\ y' \\ w \end{bmatrix} = \begin{bmatrix} a\,w\,x + b\,w\,y + l\,w \\ c\,w\,x + d\,w\,y + m\,w \\ w \end{bmatrix} = \begin{bmatrix} a & b & l \\ c & d & m \\ 0 & 0 & 1 \end{bmatrix} \begin{bmatrix} w\,x \\ w\,y \\ w \end{bmatrix} \tag{3.8}$$

$$\mathbf{P}' = \begin{bmatrix} x' \\ y' \\ z' \\ w \end{bmatrix} = \begin{bmatrix} a\,w\,x + b\,w\,y + c\,w\,z + l\,w \\ d\,w\,x + e\,w\,y + f\,w\,z + m\,w \\ g\,h\,x + h\,w\,y + k\,w\,z + n\,w \\ w \end{bmatrix} = \begin{bmatrix} a & b & c & l \\ d & e & f & m \\ g & h & k & n \\ 0 & 0 & 0 & 1 \end{bmatrix} \begin{bmatrix} w\,x \\ w\,y \\ w\,z \\ w \end{bmatrix} \tag{3.9}$$

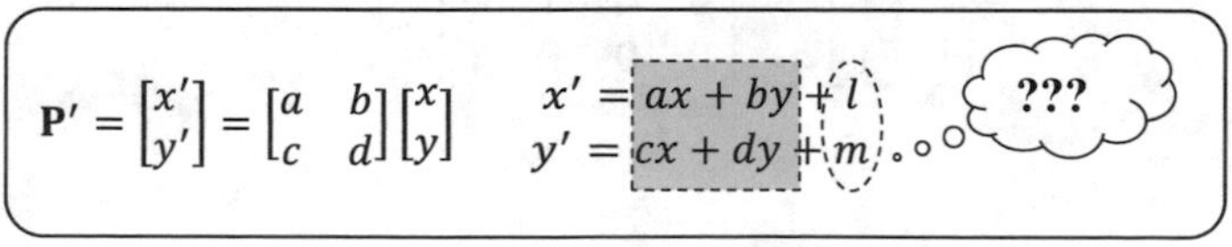

$$\mathbf{P}' = \begin{bmatrix} x' \\ y' \end{bmatrix} = \begin{bmatrix} a & b \\ c & d \end{bmatrix} \begin{bmatrix} x \\ y \end{bmatrix} \qquad \begin{array}{l} x' = ax + by + l \\ y' = cx + dy + m \end{array} \quad \text{???}$$

Fig. 3.1 Searching for a place in the matrix for translation parameters l and m

The coordinate w is called a homogeneous coordinate. It can be any real number except 0. Since the Cartesian coordinates before the transformation are multiplied by w, after the transformation the Cartesian coordinates can be obtained by mere division of the computed x', y' and z' values by w. In that case, if $w = 1$, the computed x', y' and z' are already Cartesian coordinates and the modified matrix transformation equations will implement the transformations in Eqs. (3.4) and (3.5):

$$\mathbf{P'} = \begin{bmatrix} x' \\ y' \\ 1 \end{bmatrix} = \begin{bmatrix} a\,x + b\,y + l \\ c\,x + d\,y + m \\ 1 \end{bmatrix} = \begin{bmatrix} a & b & l \\ c & d & m \\ 0 & 0 & 1 \end{bmatrix} \begin{bmatrix} x \\ y \\ 1 \end{bmatrix} \quad (3.10)$$

$$\mathbf{P'} = \begin{bmatrix} x' \\ y' \\ z' \\ 1 \end{bmatrix} = \begin{bmatrix} a\,x + b\,y + c\,z + l \\ d\,x + e\,y + f\,z + m \\ g\,x + h\,y + k\,z + n \\ 1 \end{bmatrix} = \begin{bmatrix} a & b & c & l \\ d & e & f & m \\ g & h & k & n \\ 0 & 0 & 0 & 1 \end{bmatrix} \begin{bmatrix} x \\ y \\ z \\ 1 \end{bmatrix} \quad (3.11)$$

We then simply have to add 1 as an additional coordinate to the Cartesian coordinates x, y or x, y, z, multiply them by the transformation matrix in homogeneous form, and obtain the transformed Cartesian coordinates x', y' or x', y', z'. We do not have to store the homogeneous coordinate 1 in the geometric data structures of the visualization software. It only has to be attached to the position vector at the moment of multiplying the transformation matrix by it. The last row of the transformation matrix also does not have to be remembered on the computer since it is always the same. Let's explore what the transformation matrices can do.

3.2.2 Identity Transformation

The matrices with all 0 elements except 1 in the main diagonal elements are called *identity transformations*. They do not make any changes of the coordinates, and they are equivalents of multiplication by 1 in algebra:

$$\begin{bmatrix} x \\ y \\ 1 \end{bmatrix} = \begin{bmatrix} 1 & 0 & 0 \\ 0 & 1 & 0 \\ 0 & 0 & 1 \end{bmatrix} \begin{bmatrix} x \\ y \\ 1 \end{bmatrix} \quad (3.12)$$

$$\begin{bmatrix} x \\ y \\ z \\ 1 \end{bmatrix} = \begin{bmatrix} 1 & 0 & 0 & 0 \\ 0 & 1 & 0 & 0 \\ 0 & 0 & 1 & 0 \\ 0 & 0 & 0 & 1 \end{bmatrix} \begin{bmatrix} x \\ y \\ z \\ 1 \end{bmatrix} \quad (3.13)$$

3.2.3   Scaling and Reflection

If we change the main diagonal elements to any other numbers except 0, we will define *scaling transformation*. In a matrix form it will be defined as

$$\begin{bmatrix} S_x x \\ S_y y \\ 1 \end{bmatrix} = \begin{bmatrix} S_x & 0 & 0 \\ 0 & S_y & 0 \\ 0 & 0 & 1 \end{bmatrix} \begin{bmatrix} x \\ y \\ 1 \end{bmatrix} \tag{3.14}$$

$$\begin{bmatrix} S_x x \\ S_y y \\ S_z z \\ 1 \end{bmatrix} = \begin{bmatrix} S_x & 0 & 0 & 0 \\ 0 & S_y & 0 & 0 \\ 0 & 0 & S_z & 0 \\ 0 & 0 & 0 & 1 \end{bmatrix} \begin{bmatrix} x \\ y \\ z \\ 1 \end{bmatrix} \tag{3.15}$$

The result of scaling of a square polygon is illustrated in Fig. 3.2. Notice that since every coordinate is scaled, the polygon also shifts while growing in size. If the value of the scaling coefficient is greater than 1—we have enlargement, if it is greater than zero but less than one—we have reduction.

Negative values of the main diagonal elements will define *reflection transformations*. In 2D space, there can be reflections about any of two coordinate axes and a reflection about the origin:

$$\text{Reflection about axis Y}: \quad \begin{bmatrix} -x \\ y \\ 1 \end{bmatrix} = \begin{bmatrix} -1 & 0 & 0 \\ 0 & 1 & 0 \\ 0 & 0 & 1 \end{bmatrix} \begin{bmatrix} x \\ y \\ 1 \end{bmatrix}$$

$$\text{Reflection about axis X}: \quad \begin{bmatrix} x \\ -y \\ 1 \end{bmatrix} = \begin{bmatrix} 1 & 0 & 0 \\ 0 & -1 & 0 \\ 0 & 0 & 1 \end{bmatrix} \begin{bmatrix} x \\ y \\ 1 \end{bmatrix} \tag{3.16}$$

$$\text{Reflection about the origin}: \quad \begin{bmatrix} -x \\ -y \\ 1 \end{bmatrix} = \begin{bmatrix} -1 & 0 & 0 \\ 0 & -1 & 0 \\ 0 & 0 & 1 \end{bmatrix} \begin{bmatrix} x \\ y \\ 1 \end{bmatrix}$$

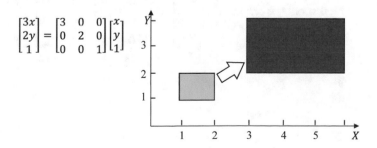

Fig. 3.2 Scaling transformation by 3 and 2 in *x* and *y* coordinates

In 3D space, we will have more variants of reflections: about the coordinate planes, about the axes, and about the origin:

$$
\text{Reflection about plane YZ:} \quad
\begin{bmatrix} -x \\ y \\ z \\ 1 \end{bmatrix}
=
\begin{bmatrix} -1 & 0 & 0 & 0 \\ 0 & 1 & 0 & 0 \\ 0 & 0 & 1 & 0 \\ 0 & 0 & 0 & 1 \end{bmatrix}
\begin{bmatrix} x \\ y \\ z \\ 1 \end{bmatrix}
\tag{3.17}
$$

$$
\text{Reflection about plane ZX:} \quad
\begin{bmatrix} x \\ -y \\ z \\ 1 \end{bmatrix}
=
\begin{bmatrix} 1 & 0 & 0 & 0 \\ 0 & -1 & 0 & 0 \\ 0 & 0 & 1 & 0 \\ 0 & 0 & 0 & 1 \end{bmatrix}
\begin{bmatrix} x \\ y \\ z \\ 1 \end{bmatrix}
$$

$$
\text{Reflection about plane XY:} \quad
\begin{bmatrix} x \\ y \\ -z \\ 1 \end{bmatrix}
=
\begin{bmatrix} 1 & 0 & 0 & 0 \\ 0 & 1 & 0 & 0 \\ 0 & 0 & -1 & 0 \\ 0 & 0 & 0 & 1 \end{bmatrix}
\begin{bmatrix} x \\ y \\ z \\ 1 \end{bmatrix}
$$

$$
\text{Reflection about axis X:} \quad
\begin{bmatrix} x \\ -y \\ -z \\ 1 \end{bmatrix}
=
\begin{bmatrix} 1 & 0 & 0 & 0 \\ 0 & -1 & 0 & 0 \\ 0 & 0 & -1 & 0 \\ 0 & 0 & 0 & 1 \end{bmatrix}
\begin{bmatrix} x \\ y \\ z \\ 1 \end{bmatrix}
$$

$$
\text{Reflection about axis Y:} \quad
\begin{bmatrix} -x \\ y \\ -z \\ 1 \end{bmatrix}
=
\begin{bmatrix} -1 & 0 & 0 & 0 \\ 0 & 1 & 0 & 0 \\ 0 & 0 & -1 & 0 \\ 0 & 0 & 0 & 1 \end{bmatrix}
\begin{bmatrix} x \\ y \\ z \\ 1 \end{bmatrix}
$$

$$
\text{Reflection about axis Z:} \quad
\begin{bmatrix} -x \\ -y \\ z \\ 1 \end{bmatrix}
=
\begin{bmatrix} -1 & 0 & 0 & 0 \\ 0 & -1 & 0 & 0 \\ 0 & 0 & 1 & 0 \\ 0 & 0 & 0 & 1 \end{bmatrix}
\begin{bmatrix} x \\ y \\ z \\ 1 \end{bmatrix}
$$

$$
\text{Reflection about the origin:} \quad
\begin{bmatrix} -x \\ -y \\ -z \\ 1 \end{bmatrix}
=
\begin{bmatrix} -1 & 0 & 0 & 0 \\ 0 & -1 & 0 & 0 \\ 0 & 0 & -1 & 0 \\ 0 & 0 & 0 & 1 \end{bmatrix}
\begin{bmatrix} x \\ y \\ z \\ 1 \end{bmatrix}
$$

3.2.4 Shear

Now, let's study the effect of the off-diagonal elements. A *shear transformation* proportional to the x-coordinate will be obtained with the following transformation where $c \neq 0$:

$$\mathbf{P}' = \begin{bmatrix} x \\ cx+y \\ 1 \end{bmatrix} = \begin{bmatrix} 1 & 0 & 0 \\ c & 1 & 0 \\ 0 & 0 & 1 \end{bmatrix} \begin{bmatrix} x \\ y \\ 1 \end{bmatrix} \tag{3.18}$$

while $b \neq 0$ yields a shear proportional to the y-coordinate (Fig. 3.3):

$$\mathbf{P}' = \begin{bmatrix} x+by \\ y \\ 1 \end{bmatrix} = \begin{bmatrix} 1 & b & 0 \\ 0 & 1 & 0 \\ 0 & 0 & 1 \end{bmatrix} \begin{bmatrix} x \\ y \end{bmatrix} \tag{3.19}$$

In the same way, a three-dimensional sheer can be defined by changing the off-diagonal elements.

3.2.5 Rotation

Let's now define *rotation transformation* in a matrix form. Let's consider a point **P** which rotates by angle θ about the origin (Fig. 3.4). For the original and transformed points **P** and **P'**, we can then write

$$\mathbf{P} = \begin{bmatrix} r\cos\phi \\ r\sin\phi \end{bmatrix} \quad \mathbf{P}' = \begin{bmatrix} r\cos(\phi+\theta) \\ r\sin(\phi+\theta) \end{bmatrix} \tag{3.20}$$

By using the sum of angles formulas, we can rewrite it as follows:

$$\mathbf{P}' = \begin{bmatrix} r(\cos\phi\,\cos\theta - \sin\phi\,\sin\theta) \\ r(\cos\phi\,\sin\theta + \sin\phi\,\cos\theta) \end{bmatrix} = \begin{bmatrix} x\cos\theta - y\sin\theta \\ x\sin\theta + y\cos\theta \end{bmatrix} \tag{3.21}$$

which yields the following coordinates of the rotated point:

$$x' = x\cos\theta - y\sin\theta \quad y' = x\sin\theta + y\cos\theta \tag{3.22}$$

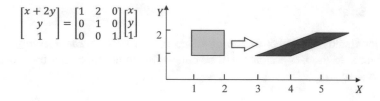

Fig. 3.3 Shear by 2 proportional to y-coordinate

Fig. 3.4 Arbitrary rotation about the origin

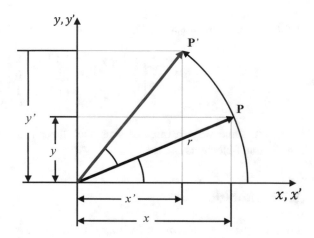

In a matrix form it becomes

$$\mathbf{P'} = \begin{bmatrix} x' \\ y' \\ 1 \end{bmatrix} = \begin{bmatrix} \cos\theta & -\sin\theta & 0 \\ \sin\theta & \cos\theta & 0 \\ 0 & 0 & 1 \end{bmatrix} \begin{bmatrix} x \\ y \\ 1 \end{bmatrix} \tag{3.23}$$

Positive rotation is defined by the right-hand rule, i.e. counter-clockwise.

While in 2D space we rotate about the origin, in 3D we can rotate about any of the three coordinate axes. It is then essential to know the direction of the axis to be able to define a positive rotation properly. We can do it by using the right-hand rule where the thumb shows the direction of the axis while the curling fingers show the positive direction of rotation about this axis.

To come up with matrices of rotation about the three principal coordinate axes, we can observe that when rotating about axis X—only y and z coordinates change (Fig. 3.5a), when rotating about axis Z—only x and y coordinate change (Fig. 3.5b), and when rotating about axis Y—only x and z coordinates change (Fig. 3.5c).

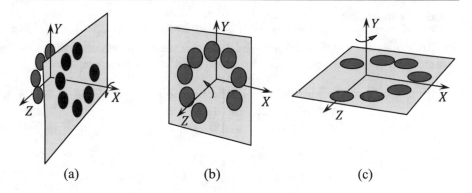

Fig. 3.5 3D rotations about **a** X-axis, **b** Z-axis and **c** Y-axis

Therefore, we can simply reuse the 3×3 2D rotation matrix by incorporating it into the 4×4 3D rotation matrix about axis X:

x does not change

$$\mathbf{P}' = \begin{bmatrix} x' \\ y' \\ z' \\ 1 \end{bmatrix} = \begin{bmatrix} 1 & 0 & 0 & 0 \\ 0 & \cos\varphi & -\sin\varphi & 0 \\ 0 & \sin\varphi & \cos\varphi & 0 \\ 0 & 0 & 0 & 1 \end{bmatrix} \begin{bmatrix} x \\ y \\ z \\ 1 \end{bmatrix} \tag{3.24}$$

The rotation is assumed positive in a right-hand sense: from axis Y to axis Z.

Similarly, rotations about axes Z and Y can be defined with the following matrices:

$$\begin{bmatrix} \cos\varphi & -\sin\varphi & 0 & 0 \\ \sin\varphi & \cos\varphi & 0 & 0 \\ 0 & 0 & 1 & 0 \\ 0 & 0 & 0 & 1 \end{bmatrix} \begin{matrix} z \text{ does not} \\ \text{change} \\ \\ y \text{ does not} \\ \text{change} \end{matrix} \begin{bmatrix} \cos\varphi & 0 & \sin\varphi & 0 \\ 0 & 1 & 0 & 0 \\ -\sin\varphi & 0 & \cos\varphi & 0 \\ 0 & 0 & 0 & 1 \end{bmatrix} \tag{3.25}$$

Note that in the matrix of rotation about axis Y the signs of the *sin* terms are reversed. This is required to maintain the positive right-hand rule convention since the original matrix has been written for X to Z rotation rather than for Z to X which we actually have.

3.2.6 Translation

Finally, *translation transformation* can be defined if we use the elements of the last column—the column which we actually added for this very transformation:

$$
\begin{bmatrix} x' \\ y' \\ 1 \end{bmatrix} = \begin{bmatrix} x+l \\ y+m \\ 1 \end{bmatrix} = \begin{bmatrix} 1 & 0 & l \\ 0 & 1 & m \\ 0 & 0 & 1 \end{bmatrix} \begin{bmatrix} x \\ y \\ 1 \end{bmatrix}
\tag{3.26}
$$

$$
\begin{bmatrix} x' \\ y' \\ z' \\ 1 \end{bmatrix} = \begin{bmatrix} x+l \\ y+m \\ z+n \\ 1 \end{bmatrix} = \begin{bmatrix} 1 & 0 & 0 & l \\ 0 & 1 & 0 & m \\ 0 & 0 & 1 & n \\ 0 & 0 & 0 & 1 \end{bmatrix} \begin{bmatrix} x \\ y \\ z \\ 1 \end{bmatrix}
\tag{3.27}
$$

So far, with the homogeneous 3×3 and 4×4 matrices we were able to implement scaling, reflection, shear, rotation, and translation transformations. Reflection can be thought of as a particular case of scaling, when negative scaling parameters are used. Shear, in fact, can be defined through rotation and scaling transformations (Fig. 3.6), though it is beyond the scope of this book to mathematically prove it.

3.3 Composition of Transformations

By this point, we know how to implement in a matrix form the three affine transformations—*scaling, rotation, and translation*—which will become basic linear transformations for constructing any other transformations that can be defined by linear functions in Eqs. (3.4) and (3.5). For doing it, we will explore the

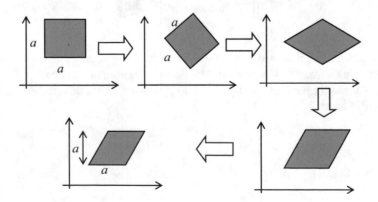

Fig. 3.6 Shear transformation as a composition of rotation and scaling transformations

opportunities which matrix algebra gives. Several consecutive affine transformations can be composed in one matrix by multiplying the respective transformation matrices.

3.3.1 Rotation About a Point

Rotation about any arbitrary point can be defined with three transformations: translation of the center of rotation toward the origin, rotation about the origin, and inverse translation (Fig. 3.7). The matrices are to be pre-multiplied, i.e., the last transformation matrix will be the left-most in the sequence of matrices.

$$
\mathbf{P'} = \begin{bmatrix} x' \\ y' \\ h \end{bmatrix} = \begin{bmatrix} 1 & 0 & l \\ 0 & 1 & m \\ 0 & 0 & 1 \end{bmatrix} \begin{bmatrix} \cos\theta & -\sin\theta & 0 \\ \sin\theta & \cos\theta & 0 \\ 0 & 0 & 1 \end{bmatrix} \begin{bmatrix} 1 & 0 & -l \\ 0 & 0 & -m \\ 0 & 0 & 1 \end{bmatrix} \begin{bmatrix} x \\ y \\ 1 \end{bmatrix}
$$

$$
= \begin{bmatrix} 1 & 0 & l \\ 0 & 1 & m \\ 0 & 0 & 1 \end{bmatrix} \begin{bmatrix} \cos\theta & -\sin\theta & -l\cos\theta + m\sin\theta \\ \sin\theta & \cos\theta & -l\sin\theta - m\cos\theta \\ 0 & 0 & 1 \end{bmatrix} \begin{bmatrix} x \\ y \\ 1 \end{bmatrix} \quad (3.28)
$$

$$
= \begin{bmatrix} \cos\theta & -\sin\theta & -l(\cos\theta - 1) + m\sin\theta \\ \sin\theta & \cos\theta & -l\sin 0 - m(\cos\theta - 1) \\ 0 & 0 & 1 \end{bmatrix} \begin{bmatrix} x \\ y \\ 1 \end{bmatrix}
$$

3.3.2 Scaling and Reflection About Points and Lines in 2D Space

Similarly, scaling relative to any arbitrary point can be implemented with two translations and one scaling transformation:

$$
\mathbf{P'} = \begin{bmatrix} x' \\ y' \\ h \end{bmatrix} = \begin{bmatrix} 1 & 0 & l \\ 0 & 1 & m \\ 0 & 0 & 1 \end{bmatrix} \begin{bmatrix} s_x & 0 & 0 \\ 0 & s_y & 0 \\ 0 & 0 & 1 \end{bmatrix} \begin{bmatrix} 1 & 0 & -l \\ 0 & 0 & -m \\ 0 & 0 & 1 \end{bmatrix} \begin{bmatrix} x \\ y \\ 1 \end{bmatrix}
$$

$$
= \begin{bmatrix} s_x & 0 & l(1 - s_x) \\ 0 & s_y & m(1 - s_y) \\ 0 & 0 & 1 \end{bmatrix} \begin{bmatrix} x \\ y \\ 1 \end{bmatrix} \quad (3.29)
$$

A more complex example of reflection about an arbitrary straight line is illustrated in Fig. 3.8. In this example, it takes five matrices to be composed into one performing the transformation: $\mathbf{T} = \mathbf{T}_{trans}^{-1} \mathbf{T}_{rot}^{-1} \mathbf{T}_{ref} \mathbf{T}_{rot} \mathbf{T}_{trans}$.

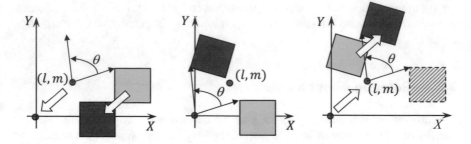

Fig. 3.7 Rotation about an arbitrary point

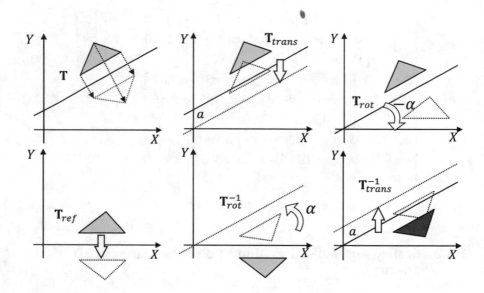

Fig. 3.8 Reflection about an arbitrary straight line

$$
\mathbf{T}_{trans} = \begin{bmatrix} 1 & 0 & 0 \\ 0 & 1 & -a \\ 0 & 0 & 1 \end{bmatrix} \quad
\mathbf{T}_{rot} = \begin{bmatrix} \cos(-\alpha) & -\sin(-\alpha) & 0 \\ \sin(-\alpha) & \cos(-\alpha) & 0 \\ 0 & 0 & 1 \end{bmatrix}
$$

$$
\mathbf{T}_{ref} = \begin{bmatrix} 1 & 0 & 0 \\ 0 & -1 & 0 \\ 0 & 0 & 1 \end{bmatrix} \quad
\mathbf{T}_{rot}^{-1} = \begin{bmatrix} \cos(\alpha) & -\sin(\alpha) & 0 \\ \sin(\alpha) & \cos(\alpha) & 0 \\ 0 & 0 & 1 \end{bmatrix} \quad
\mathbf{T}_{trans}^{-1} = \begin{bmatrix} 1 & 0 & 0 \\ 0 & 1 & a \\ 0 & 0 & 1 \end{bmatrix}
$$

$$(3.30)$$

3.3.3 Deriving Matrix of an Arbitrary 2D Affine Transformation

Let's consider a converse problem which has to be solved when we perform affine transformations of one coordinate system into another. With reference to Fig. 3.9, we will need to derive an affine transformation matrix which will transform shape A into shape B.

The general matrix transformation Eq. (3.8) has 6 unknown variables a, b, c, d, l, m and, therefore, it requires 6 equations which in turn require 6 coordinates x, y that can be obtained from 3 independent points. We will select the following three points with coordinates x, y: (0, 0), (4, 0), (0, 2). The respective transformed coordinates x', y' are (8, 8), (6, 4), (12, 8). Substituting these coordinates into the transformation equations yields

$(0, 0) \rightarrow (8, 8)$ which yields $l = 8, m = 8$
$(4, 0) \rightarrow (6, 4)$ which yields $6 = 4a + 8 \rightarrow a = -0.5$ $4 = 4d + 8 \rightarrow c = -1$
$(0, 2) \rightarrow (12, 8)$ which yields $12 = 2b + 8 \rightarrow b = 2$ $8 = 2e + 8 \rightarrow d = 0$

$$\text{The resulting matrix follows: } \mathbf{T} = \begin{bmatrix} -0.5 & 2 & 8 \\ -1 & 0 & 8 \\ 0 & 0 & 1 \end{bmatrix}$$

3.3.4 Rotation About an Axis

For 3D rotation about an axis parallel to any of three coordinate axes, we will use a composition of transformations similar to the one used in 2D for rotation about an arbitrary point. We will translate the object until the axis of rotation is coincident with the coordinate axis in the same direction, then rotate about this axis, and finally translate the transformed object back to where it is supposed to be placed:

$$\mathbf{P}' = \mathbf{T}_{trans(l,m,n)} \mathbf{T}_{rot(asix\ X, Y\ or\ Z)} \mathbf{T}_{trans(-l,-m,-n)} \mathbf{P} \tag{3.31}$$

Fig. 3.9 Affine transformation of a complex object

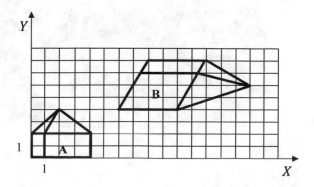

Matrix implementation of the rotation about an arbitrary axis passing through the origin is not so straightforward, although the same method of composing matrix transformation is to be used here. The goal of these transformations is to make an axis of rotation coincident with one of the coordinate axes. Then, we will rotate about this axis. Finally, by applying inverse transformations, we have to restore the original position of the axis of rotation in space together by transforming to the relevant position the result of rotation. To bring the axis of rotation to one of the coordinate axes, maximum of two rotations about the other two coordinate axes are required (Fig. 3.10).

Thus, to eventually rotate about axis Y, we could choose to rotate first about axis Z and then about axis X. We need to determine the angles of these two rotations, or rather cos and sin terms of the respective rotation matrices since they are used in the matrix of rotation. The axis of rotation is usually defined by coordinates of a point on it, or a position vector $P = [x \quad y \quad z]$. We also can calculate so-called direction cosines $c_x, c_y,$ and c_z of this position vector that are cosines of angles $\alpha, \beta,$ and γ (Fig. 3.11):

$$c_x = \cos \alpha = \frac{x}{\sqrt{x^2+y^2+z^2}} \qquad c_y = \cos \beta = \frac{y}{\sqrt{x^2+y^2+z^2}}$$
$$c_z = \cos \gamma = \frac{z}{\sqrt{x^2+y^2+z^2}} \tag{3.32}$$

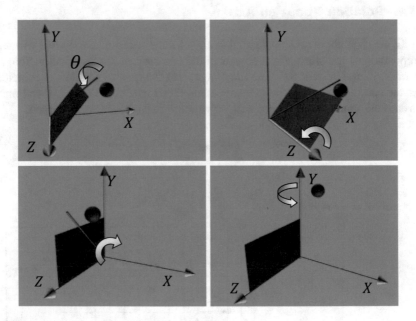

Fig. 3.10 Rotation of the axis of rotation to the vertical coordinate axis

Note that for the unit position vector P, its magnitude $\sqrt{x^2 + y^2 + z^2} = 1$, and therefore its direction cosines are equal to its coordinates $c_x = x$, $c_y = y$, $c_z = z$. Therefore, in our further considerations, we will calculate the rotation matrices for the unit vector using only direction cosines which are the same for any point on the axis.

First, we need to rotate vector P about axis Z towards plane YZ so that eventually it will be located in this plane. For the respective matrix of rotation, we need to know $\cos \alpha$ and $\sin \alpha$ of the rotation angle α which can easily be obtained on plane XY (Fig. 3.12). This yields the first rotation matrix:

$$\mathbf{T}_z = \begin{bmatrix} \cos\alpha & -\sin\alpha & 0 & 0 \\ \sin\alpha & \cos\alpha & 0 & 0 \\ 0 & 0 & 1 & 0 \\ 0 & 0 & 0 & 1 \end{bmatrix} = \begin{bmatrix} \frac{c_y}{\sqrt{c_x^2 + c_y^2}} & -\frac{c_x}{\sqrt{c_x^2 + c_y^2}} & 0 & 0 \\ \frac{c_x}{\sqrt{c_x^2 + c_y^2}} & \frac{c_y}{\sqrt{c_x^2 + c_y^2}} & 0 & 0 \\ 0 & 0 & 1 & 0 \\ 0 & 0 & 0 & 1 \end{bmatrix} \qquad (3.33)$$

Next, we have to rotate vector P about axis X towards axis Y. The parameters of the respective matrix of rotation can be obtained as it is illustrated in Fig. 3.13.

Finally, we rotate by angle θ about axis Y and perform two inverse rotations to bring the result back to its actual location in 3D space (Fig. 3.14).

Thus, a composition of five rotation matrices yields the final matrix of rotation about an arbitrary axis passing through the origin. Note that we could achieve exactly the same result by rotating the axis of rotation in a different order, e.g., $X \rightarrow Z \rightarrow Y$, or $Z \rightarrow Y \rightarrow X$, or $Y \rightarrow Z \rightarrow X$, etc. If the axis of rotation is not passing through the origin, we have to add two translation transformations making up to 7 transformation matrices involved in this rotation.

The number of matrices will be smaller if the axis of rotation is parallel to one of the coordinate planes. Let's consider an example of rotation about an axis passing through some point P while being parallel to plane ZX (Fig. 3.15).

First, we should translate this axis so that it will pass through the origin. This can be done with the following matrix and the respective inverse transformation matrix:

Fig. 3.11 Direction cosines of vector $P = \begin{bmatrix} x & y & z \end{bmatrix}$

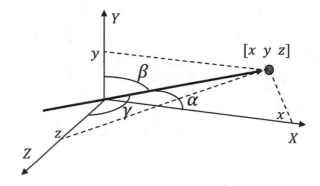

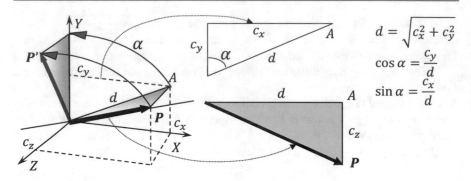

Fig. 3.12 Calculating the first matrix of rotation about axis Z

$$\cos\beta = \frac{d}{|\boldsymbol{P}|} = d = \sqrt{c_x^2 + c_y^2} \quad \sin\beta = \frac{c_z}{|\boldsymbol{P}|} = c_z$$

$$\mathbf{T}_x = \begin{bmatrix} 1 & 0 & 0 & 0 \\ 0 & \cos(-\beta) & -\sin(-\beta) & 0 \\ 0 & \sin(-\beta) & \cos(-\beta) & 0 \\ 0 & 0 & 0 & 1 \end{bmatrix}$$

$$= \begin{bmatrix} 1 & 0 & 0 & 0 \\ 0 & \sqrt{c_x^2 + c_y^2} & c_z & 0 \\ 0 & -c_z & \sqrt{c_x^2 + c_y^2} & 0 \\ 0 & 0 & 0 & 1 \end{bmatrix}$$

Fig. 3.13 Calculating the second matrix of rotation about axis X

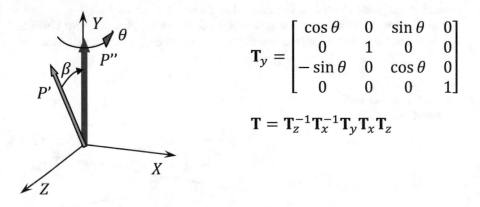

$$\mathbf{T}_y = \begin{bmatrix} \cos\theta & 0 & \sin\theta & 0 \\ 0 & 1 & 0 & 0 \\ -\sin\theta & 0 & \cos\theta & 0 \\ 0 & 0 & 0 & 1 \end{bmatrix}$$

$$\mathbf{T} = \mathbf{T}_z^{-1}\mathbf{T}_x^{-1}\mathbf{T}_y\mathbf{T}_x\mathbf{T}_z$$

Fig. 3.14 Final rotation about axis Y and the composed transformation **T**

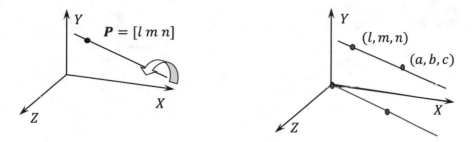

Fig. 3.15 An example of rotating about an arbitrary axis

$$\mathbf{T}_{trans} = \begin{bmatrix} 1 & 0 & 0 & -l \\ 0 & 1 & 0 & -m \\ 0 & 0 & 1 & -n \\ 0 & 0 & 0 & 1 \end{bmatrix} \quad \mathbf{T}_{trans}^{-1} = \begin{bmatrix} 1 & 0 & 0 & l \\ 0 & 1 & 0 & m \\ 0 & 0 & 1 & n \\ 0 & 0 & 0 & 1 \end{bmatrix} \quad (3.34)$$

The direction cosines of the axis will be calculated as follows:

$$c_x = \frac{a-l}{\sqrt{(a-l)^2 + (b-m)^2 + (c-n)^2}}$$

$$c_z = \frac{c-n}{\sqrt{(a-l)^2 + (b-m)^2 + (c-n)^2}} \quad (3.35)$$

Only one rotation is required to make an axis coincident with axis X or Z. We choose axis X, and the necessary matrices of rotation will be the following:

$$\mathbf{T}_y = \begin{bmatrix} c_x & 0 & c_z & 0 \\ 0 & 1 & 0 & 0 \\ -c_z & 0 & c_x & 0 \\ 0 & 0 & 0 & 1 \end{bmatrix} \quad \mathbf{T}_x = \begin{bmatrix} 1 & 0 & 0 & 0 \\ 0 & \cos\beta & -\sin\beta & 0 \\ 0 & \sin\beta & \cos\beta & 0 \\ 0 & 0 & 0 & 1 \end{bmatrix}$$

$$\mathbf{T}_y^{-1} = \begin{bmatrix} c_x & 0 & -c_z & 0 \\ 0 & 1 & 0 & 0 \\ c_z & 0 & c_x & 0 \\ 0 & 0 & 0 & 1 \end{bmatrix} \quad (3.36)$$

Finally, the whole sequence of transformations can be written as

$$\mathbf{T} = \mathbf{T}_{trans}^{-1} \mathbf{T}_y^{-1} \mathbf{T}_x \mathbf{T}_y \mathbf{T}_{trans} \quad (3.37)$$

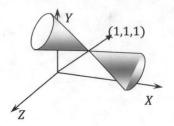

3.3.5   Reflections About Any Point, Axis, or Plane in 3D Space

Like in 2D space, besides reflection about the origin, we can also define reflection about the principal coordinate planes, about the principal coordinate axis, as well as about arbitrary located planes and axes. By using composition of transformations, we can derive matrices for all these transformations. Let's consider an example (Fig. 3.16) of reflection about a straight line passing through the origin and point P with coordinates $(1, 1, 1)$.

We aim to transform this line so that it will be coincident with axis Y. Like in case of 3D rotation, it will require two rotation transformations about Z and X axes. First, we calculate the direction cosines:

$$c_x = \frac{1}{\sqrt{1+1+1}} = \frac{1}{\sqrt{3}} \quad c_y = \frac{1}{\sqrt{3}} \quad c_z = \frac{1}{\sqrt{3}} \tag{3.38}$$

Next, with reference to matrices obtained for 3D rotations in (5.2) the following holds:

$$\mathbf{T}_z = \begin{bmatrix} \frac{c_y}{\sqrt{c_x^2+c_y^2}} & -\frac{c_x}{\sqrt{c_x^2+c_y^2}} & 0 & 0 \\ \frac{c_x}{\sqrt{c_x^2+c_y^2}} & \frac{c_y}{\sqrt{c_x^2+c_y^2}} & 0 & 0 \\ 0 & 0 & 1 & 0 \\ 0 & 0 & 0 & 1 \end{bmatrix} = \begin{bmatrix} \frac{1}{\sqrt{2}} & -\frac{1}{\sqrt{2}} & 0 & 0 \\ \frac{1}{\sqrt{2}} & \frac{1}{\sqrt{2}} & 0 & 0 \\ 0 & 0 & 1 & 0 \\ 0 & 0 & 0 & 1 \end{bmatrix}$$

$$\mathbf{T}_x = \begin{bmatrix} 1 & 0 & 0 & 0 \\ 0 & \sqrt{c_x^2+c_y^2} & -c_z & 0 \\ 0 & c_z & \sqrt{c_x^2+c_y^2} & 0 \\ 0 & 0 & 0 & 1 \end{bmatrix} = \begin{bmatrix} 1 & 0 & 0 & 0 \\ 0 & \sqrt{\frac{2}{3}} & -\frac{1}{\sqrt{3}} & 0 \\ 0 & \frac{1}{\sqrt{3}} & \sqrt{\frac{2}{3}} & 0 \\ 0 & 0 & 0 & 1 \end{bmatrix} \tag{3.39}$$

$$\mathbf{T}_{ref(y)} = \begin{bmatrix} -1 & 0 & 0 & 0 \\ 0 & 1 & 0 & 0 \\ 0 & 0 & -1 & 0 \\ 0 & 0 & 0 & 1 \end{bmatrix}$$

The final matrix will be obtained as the following composition of five transformations:

$$\mathbf{T} = \mathbf{T}_z^{-1}\mathbf{T}_x^{-1}\mathbf{T}_{ref(y)}\mathbf{T}_x\mathbf{T}_z \tag{3.40}$$

The resulting matrix transformation equation will be as follows:

$$\begin{bmatrix} x' \\ y' \\ z' \\ 1 \end{bmatrix} = \begin{bmatrix} \frac{1}{\sqrt{2}} & \frac{1}{\sqrt{2}} & 0 & 0 \\ -\frac{1}{\sqrt{2}} & \frac{1}{\sqrt{2}} & 0 & 0 \\ 0 & 0 & 1 & 0 \\ 0 & 0 & 0 & 1 \end{bmatrix} \begin{bmatrix} 1 & 0 & 0 & 0 \\ 0 & \sqrt{\frac{2}{3}} & \frac{1}{\sqrt{3}} & 0 \\ 0 & -\frac{1}{\sqrt{3}} & \sqrt{\frac{2}{3}} & 0 \\ 0 & 0 & 0 & 1 \end{bmatrix} \begin{bmatrix} -1 & 0 & 0 & 0 \\ 0 & 1 & 0 & 0 \\ 0 & 0 & -1 & 0 \\ 0 & 0 & 0 & 1 \end{bmatrix}$$
$$\begin{bmatrix} 1 & 0 & 0 & 0 \\ 0 & \sqrt{\frac{2}{3}} & -\frac{1}{\sqrt{3}} & 0 \\ 0 & \frac{1}{\sqrt{3}} & \sqrt{\frac{2}{3}} & 0 \\ 0 & 0 & 0 & 1 \end{bmatrix} \begin{bmatrix} \frac{1}{\sqrt{2}} & -\frac{1}{\sqrt{2}} & 0 & 0 \\ \frac{1}{\sqrt{2}} & \frac{1}{\sqrt{2}} & 0 & 0 \\ 0 & 0 & 1 & 0 \\ 0 & 0 & 0 & 1 \end{bmatrix} \begin{bmatrix} x \\ y \\ z \\ 1 \end{bmatrix} \tag{3.41}$$

Let's consider one more example of reflection about a plane as displayed in Fig. 3.17.

This can be done by rotating the plane to make it coincident with one of the coordinate planes (e.g., YZ), followed by reflection about this plane, and the inverse rotation:

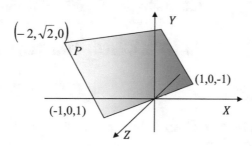

Fig. 3.17 Reflection about a plane

$$\mathbf{Rot}(Y, -45°)\mathbf{Rot}(Z, 45°)\mathbf{Ref}(YZ)\mathbf{Rot}(Z, -45°)\mathbf{Rot}(Y, 45°)$$

$$\begin{bmatrix} 0.7 & 0 & -0.7 & 0 \\ 0 & 1 & 0 & 0 \\ 0.7 & 0 & 0.7 & 0 \\ 0 & 0 & 0 & 1 \end{bmatrix} \begin{bmatrix} 0.7 & -0.7 & 0 & 0 \\ 0.7 & 0.7 & 0 & 0 \\ 0 & 0 & 1 & 0 \\ 0 & 0 & 0 & 1 \end{bmatrix} \begin{bmatrix} -1 & 0 & 0 & 0 \\ 0 & 1 & 0 & 0 \\ 0 & 0 & 1 & 0 \\ 0 & 0 & 0 & 1 \end{bmatrix}$$

$$\begin{bmatrix} 0.7 & 0.7 & 0 & 0 \\ -0.7 & 0.7 & 0 & 0 \\ 0 & 0 & 1 & 0 \\ 0 & 0 & 0 & 1 \end{bmatrix} \begin{bmatrix} 0.7 & 0 & 0.7 & 0 \\ 0 & 1 & 0 & 0 \\ -0.7 & 0 & 0.7 & 0 \\ 0 & 0 & 0 & 1 \end{bmatrix} \tag{3.42}$$

Actually, there is very little difference in obtaining a matrix for rotation about any axis and reflection about a straight line or a plane. For rotation, we move the axis of rotation to coincide with one of the coordinate axes which is done by translation and up to two rotation transformations and then we rotate about this coordinate axis. For reflection about a straight line, we move the line to coincide with one of the coordinate axes also by translation and up to two rotation transformations and then we reflect about this coordinate axis. For reflection about a plane, we move the plane so that its normal vector coincides with one of the coordinate axes also by using translation and up to two rotation transformations and then we reflect about the coordinate plane orthogonal to this coordinate axis. So, the difference is only in one central matrix:

Rotation: $$\mathbf{T} = \mathbf{T}_{trans}^{-1}\mathbf{T}_{rot1}^{-1}\mathbf{T}_{rot2}^{-1}\boxed{\mathbf{T}_{rot3}}\mathbf{T}_{rot2}\mathbf{T}_{rot1}\mathbf{T}_{trans} \tag{3.43}$$

Reflection: $$\mathbf{T} = \mathbf{T}_{trans}^{-1}\mathbf{T}_{rot1}^{-1}\mathbf{T}_{rot2}^{-1}\boxed{\mathbf{T}_{ref}}\mathbf{T}_{rot2}\mathbf{T}_{rot1}\mathbf{T}_{trans} \tag{3.44}$$

3.3.6 Deriving Matrix of an Arbitrary 3D Affine Transformation

Like in 2D, we may also need to derive a matrix for a complex affine transformation which transforms one object into another, or one coordinate system into another. For example, an object with coordinates (0, 0, 0), (1, 0, 0), (1, 1, 0), (0, 1, 0), (0, 0, 1), (1, 0, 1), (1, 1, 1), (0, 1, 1) is transformed by some complex affine transformation so that the respective transformed coordinates are: (4, 5, 6), (5, 7, 9), (7, 10, 13), (6, 8, 10), (7, 9, 11), (8, 11, 14), (10, 14, 18), (9, 12, 15). Since the general affine transformation matrix has 12 unknown values, we need four equations that requires us to select four non-collinear points and write transformation equations for them:

$$(0,0,0) \rightarrow (4,5,6)$$
$$(1,0,0) \rightarrow (5,7,9)$$
$$(0,1,0) \rightarrow (6,8,10)$$
$$(0,0,1) \rightarrow (7,9,11)$$

A matrix transformation equation is $\begin{bmatrix} x' \\ y' \\ z' \\ 1 \end{bmatrix} = \begin{bmatrix} a & d & g & l \\ b & e & h & m \\ c & f & k & n \\ 0 & 0 & 0 & 1 \end{bmatrix} \begin{bmatrix} x \\ y \\ z \\ 1 \end{bmatrix}$.

Substituting the coordinates yields:

$$4 = l \qquad\qquad 5 = m \qquad\qquad\qquad 6 = n$$
$$5 = a+4 \rightarrow a = 1 \quad 7 = b+5 \rightarrow b = 2 \quad 9 = c+6 \rightarrow c = 3$$
$$6 = d+4 \rightarrow d = 2 \quad 8 = e+5 \rightarrow e = 3 \quad 10 = f+6 \rightarrow f = 4$$
$$7 = g+4 \rightarrow g = 3 \quad 9 = h+5 \rightarrow h = 4 \quad 11 = k+6 \rightarrow k = 5$$

The final matrix is

$$\begin{bmatrix} 1 & 2 & 3 & 4 \\ 2 & 3 & 4 & 5 \\ 3 & 4 & 5 & 6 \\ 0 & 0 & 0 & 1 \end{bmatrix}$$

3.3.7 Matrix Algebra Laws

It is important to highlight that the operations on matrices are not commutative. That is, in general, for two matrices $\mathbf{T}_1$ and $\mathbf{T}_2$, their product $\mathbf{T}_1\mathbf{T}_2$ is not equal to $\mathbf{T}_2\mathbf{T}_1$. Thus, the order of multiplications is important. For example, rotation followed by scaling will generally have a different result from that of scaling followed by rotation.

Matrix operations follow the first and second *distributive laws*:

$$\mathbf{T}_1(\mathbf{T}_2 + \mathbf{T}_3) = \mathbf{T}_1\mathbf{T}_2 + \mathbf{T}_1\mathbf{T}_3 \quad \text{and} \quad (\mathbf{T}_1 + \mathbf{T}_2)\mathbf{T}_3 = \mathbf{T}_1\mathbf{T}_3 + \mathbf{T}_2\mathbf{T}_3 \qquad (3.45)$$

Also, the *associative law* applies:

$$\mathbf{T}_1(\mathbf{T}_2\mathbf{T}_3) = (\mathbf{T}_1\mathbf{T}_2)\mathbf{T}_3 \qquad (3.46)$$

Sometimes, position vectors of points are represented in row vector form $[x \quad y \quad 1]$ rather than in column form. In that case, we still can apply the matrices derived in this chapter as follows:

$$\mathbf{P'} = [x' \quad y' \quad 1] = [x \quad y \quad 1] \begin{bmatrix} a & c & 0 \\ b & d & 0 \\ m & n & 1 \end{bmatrix}$$
$$= [x \quad y \quad 1]\mathbf{T}_1^T\mathbf{T}_2^T\mathbf{T}_3^T\ldots\mathbf{T}_n^T \qquad (3.47)$$

where $\mathbf{T}_1^T, \mathbf{T}_2^T, \mathbf{T}_3^T, \ldots, \mathbf{T}_n^T$ are transposed matrices from the matrix equation for column representation of a position vector:

$$\mathbf{P'} = \begin{bmatrix} x' \\ y' \\ 1 \end{bmatrix} = \begin{bmatrix} a & b & m \\ c & d & n \\ 0 & 0 & 1 \end{bmatrix} \begin{bmatrix} x \\ y \\ 1 \end{bmatrix} = \mathbf{T}_n\ldots\mathbf{T}_3\mathbf{T}_2\mathbf{T}_1 \begin{bmatrix} x \\ y \\ 1 \end{bmatrix} \qquad (3.48)$$

Notice that for column representation, the matrices are listed backwards in contrast to row representation. For example, rotation about an arbitrary point will be defined for row represented position vector as follows:

$$\mathbf{P'} = [x' \quad y' \quad 1] = [x \quad y \quad 1]\mathbf{T}_{trans(-m,-n)}\mathbf{T}_{rot(\theta)}\mathbf{T}_{trans(m,n)} \qquad (3.49)$$

3.3.8 Definition of Sweeping by Transformation Matrices

Now, that we know how to properly combine different basic transformation to create any complex affine transformation, we can now define surfaces by making any complex transformation of non-planar curves as

$$\begin{bmatrix} x(u,v) \\ y(u,v) \\ z(u,v) \\ 1 \end{bmatrix} = \begin{bmatrix} a(v) & b(v) & c(v) & l(v) \\ d(v) & e(v) & f(v) & m(v) \\ g(v) & h(v) & k(v) & n(v) \\ 0 & 0 & 0 & 1 \end{bmatrix} \begin{bmatrix} x(u) \\ y(u) \\ z(u) \\ 1 \end{bmatrix} \qquad (3.50)$$

where $x(u), y(u)$, and $z(u)$ are parametric definitions of a curve.

Let's define a surface by rotating a 3D parametric curve defined by

$$x(u) = 0.5\cos(2\pi u)$$
$$y(u) = 0.5\sin(2\pi u) \qquad (3.51)$$
$$z(u) = 0.5\cos(2\pi u)\sin(2\pi u)$$

by angle $-\pi$ about an axis passing through the points with coordinates $(1, 0, 1)$ and $(1, 1, 1)$ (Fig. 3.18a, b).

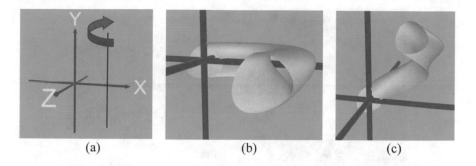

(a)   (b)   (c)

Fig. 3.18 **a** A 3D parametric curve and **b** a surface created by its rotational sweeping

We have to do the following transformations:

$$
\begin{bmatrix} x(u,v) \\ y(u,v) \\ z(u,v) \\ 1 \end{bmatrix} = \begin{bmatrix} 1 & 0 & 0 & 1 \\ 0 & 1 & 0 & 0 \\ 0 & 0 & 1 & 1 \\ 0 & 0 & 0 & 1 \end{bmatrix} \begin{bmatrix} \cos(\pi v) & 0 & \sin(\pi v) & 0 \\ 0 & 1 & 0 & 0 \\ \sin(\pi v) & 0 & \cos(\pi v) & 0 \\ 0 & 0 & 0 & 1 \end{bmatrix} \begin{bmatrix} 1 & 0 & 0 & 1 \\ 0 & 1 & 0 & 0 \\ 0 & 0 & 1 & -1 \\ 0 & 0 & 0 & 1 \end{bmatrix} \begin{bmatrix} x(u) \\ y(u) \\ z(u) \\ 1 \end{bmatrix}
$$

$$
= \begin{bmatrix} x(u)\cos(\pi v) - z(u)\sin(\pi v) + 1 - \cos(\pi v) + \sin(\pi v) \\ y(u) \\ x(u)\sin(\pi v) + z(u)\cos(\pi v) + 1 - \sin(\pi v) - \cos(\pi v) \\ 1 \end{bmatrix}
$$

$$(3.52)$$

If we want to add a translational component that has to follow each increment of rotation, for example, by 3 units along axis Y (Fig. 3.18c), it will be done as

$$
\begin{bmatrix} 1 & 0 & 0 & 0 \\ 0 & 1 & 0 & 3v \\ 0 & 0 & 1 & 0 \\ 0 & 0 & 0 & 1 \end{bmatrix} \begin{bmatrix} x(u,v) \\ y(u,v) \\ z(u,v) \end{bmatrix} = \begin{bmatrix} x(u)\cos(\pi v) - z(u)\sin(\pi v) + 1 - \cos(\pi v) + \sin(\pi v) \\ y(u) + 3v \\ x(u)\sin(\pi v) + z(u)\cos(\pi v) + 1 - \sin(\pi v) - \cos(\pi v) \\ 1 \end{bmatrix}
$$

$$(3.53)$$

3.4 Projection Transformations

3D shapes are defined with three coordinates while images are defined with two coordinates. To be able to visualize a 3D shape, we need to perform a coordinate mapping from the 3D World Coordinate System (WCS) to the 2D Device Coordinate System (DCS). This mapping will consist of the Viewing Coordinate Transformation from the WCS to the Viewer's Coordinate System (VCS), followed by the Projection Transformation [4].

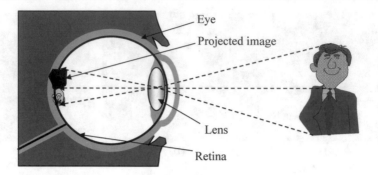

Fig. 3.19 How the eye works

When seeing the world around us, images are projected onto the eye's retina like onto a projection screen (Fig. 3.19).

3.4.1   Implementations of Projection Transformations

A straightforward implementation of human vision with a computer could be to cast rays from one or several light sources and trace them with all their possible reflections and refractions until they reach the observer's eye. However, it is impossible since there will be only a very small portion of rays from the light source eventually reaching the eye. As such, an inverse method, called *ray tracing*, is commonly used in computer graphics (Fig. 3.20). Here, we trace the rays backwards: from the eye to the objects in the scene and back to the light source. Then, a projection plane containing a pixel matrix is to be placed between the observer and the scene. For each pixel, a ray is to be cast from the observer's position through the pixel towards the scene. Next, the ray is to be intersected with the shapes in the scene. Finally, the reflected and refracted rays are to be calculated and tested on whether they come back to the light source. Those rays that eventually reach the light source will contribute to the color value of the respective pixel on the screen, while those obstructed by objects will not. The number of rays to be cast back from the eye is equal to the number of pixels in the graphics viewport. If we consider only shapes defined by points, then this procedure is to be performed for each point in the scene to obtain its 2D projection on the projection plane.

Depending on the location of the observer and the projection plane, the projection of the same object will look different. For example, if a cube is to be projected with an observer located in front of it, the projection may look like those in Fig. 3.21 depending on the distance of the observer from the projection plane. For the distances that are large and can be considered as infinite, the cube will project into a square.

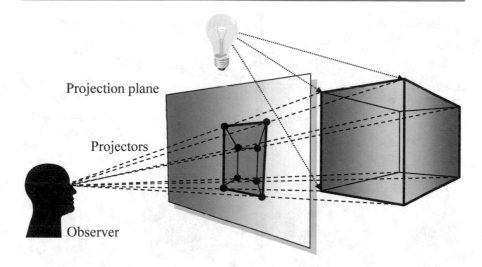

Fig. 3.20 Ray tracing

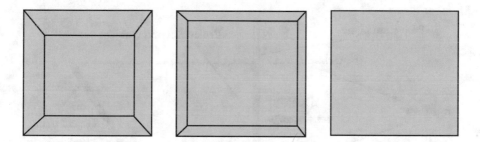

Fig. 3.21 Projections of a cube for three different locations of the observer

3.4.2 Classifications of Projection Transformations

All the projection transformations can be classified into two groups: *central projections* and *parallel projection*. For central projections, the rays (*projectors*) cast from the observer's position are non-parallel lines, and therefore distortions of size will occur in projections as illustrated in Fig. 3.22.

Parallel projections are created when the observer is so far away from the projection plane, that this distance may be considered as an infinite distance, and therefore, all the projectors cast from the observer's position will be parallel lines. If the Direction of Projection Vector (DOP) is orthogonal to the projection plane, then the projection is called *parallel orthographic projection* (Fig. 3.23).

Parallel projections do not distort sizes and can be used for measurements. Therefore, they are commonly used in engineering drawing as *Top, Front, and Side*

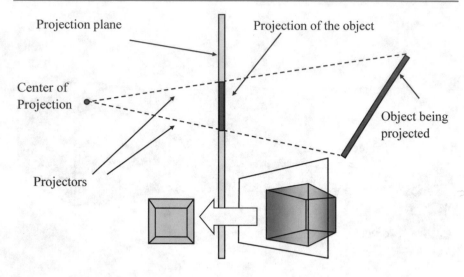

Fig. 3.22 Central projection

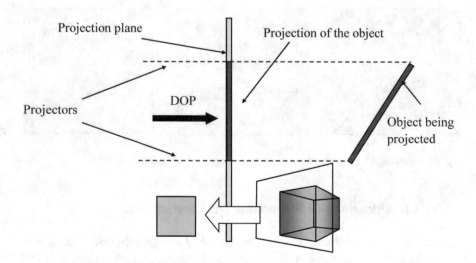

Fig. 3.23 Parallel orthographic projection

Projections of the objects being designed (Fig. 3.24). Parallel lines always remain parallel in parallel projections, which may look confusing for us since we are accustomed to see 3D objects with perspective distortions.

If the projection plane is not orthogonal to the principal axes, then we can see at once at least three adjacent faces of an object. These projections are called *axonometric projections* (Fig. 3.25). They can be implemented as a composition of

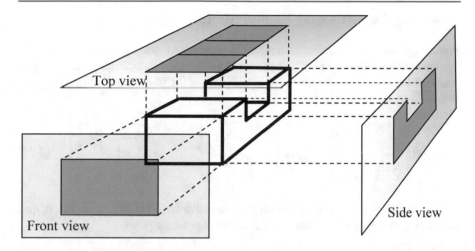

Fig. 3.24 Three main parallel orthogonal projections

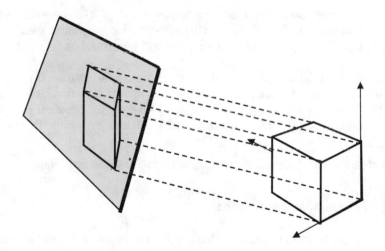

Fig. 3.25 Parallel axonometric projection

two rotations about vertical and horizontal axes, followed by a parallel orthographic projection. Axonometric projections are used in engineering drawing as auxiliary views to provide more information about the shape.

3.4.3 Parallel Orthographic Projections

Front, top, and side parallel orthographic projections are the easiest for matrix implementation. They can be interpreted as projections onto three principal coordinate planes which can be treated as setting one of the three coordinates to zero. In matrix form, it can be defined as follows:

$$\mathbf{T}_{xy} = \begin{bmatrix} 1 & 0 & 0 & 0 \\ 0 & 1 & 0 & 0 \\ 0 & 0 & 0 & 0 \\ 0 & 0 & 0 & 1 \end{bmatrix} \quad \mathbf{T}_{zx} = \begin{bmatrix} 1 & 0 & 0 & 0 \\ 0 & 0 & 0 & 0 \\ 0 & 0 & 1 & 0 \\ 0 & 0 & 0 & 1 \end{bmatrix} \quad \mathbf{T}_{yz} = \begin{bmatrix} 0 & 0 & 0 & 0 \\ 0 & 1 & 0 & 0 \\ 0 & 0 & 1 & 0 \\ 0 & 0 & 0 & 1 \end{bmatrix} \quad (3.54)$$

The zero rows in these matrices set the respective coordinates z, y, or x to 0, thus performing the projection onto the respective coordinate plane.

3.4.4 Axonometric Parallel Projections

Axonometric projections can easily be implemented as a composition of two rotations followed by one of the parallel orthographic projection matrices. For example, a projection matrix onto plane $z = 0$ will be defined as follows:

$$\begin{bmatrix} 1 & 0 & 0 & 0 \\ 0 & 1 & 0 & 0 \\ 0 & 0 & 0 & 0 \\ 0 & 0 & 0 & 1 \end{bmatrix} \begin{bmatrix} 1 & 0 & 0 & 0 \\ 0 & \cos\beta & -\sin\beta & 0 \\ 0 & \sin\beta & \cos\beta & 0 \\ 0 & 0 & 0 & 1 \end{bmatrix} \begin{bmatrix} \cos\alpha & 0 & \sin\alpha & 0 \\ 0 & 1 & 0 & 0 \\ -\sin\alpha & 0 & \cos\alpha & 0 \\ 0 & 0 & 0 & 1 \end{bmatrix} \quad (3.55)$$

where angles α and β depend on the particular axonometric projection.

3.4.5 Perspective Projections

If we are going to make an orthographic parallel projection of a street with houses on both sides, it will look like it is sketched in Fig. 3.26a, while its central projection in Fig. 3.26b will provide a more realistic view with perspective distortions. Lines, which were originally parallel, will converge at so-called *vanishing points*. All these vanishing points are located at the imaginary *horizon line* which is, in turn, located at the eye level of the observer. The observer can see the top of an object if it is below the eye level and, respectively, below the horizon line. If an object is above the eye level, that is above the horizon line, the observer cannot see its top. This systematic method of building perspective projections was invented in the early 1400s by Filippo Brunelleschi, published by Leon Battista Alberti in 1435 and further studied and developed by other famous Renaissance artists.

To come up with a matrix implementation of such *perspective projection transformations*, let's consider how a single point can be projected. In Fig. 3.27,

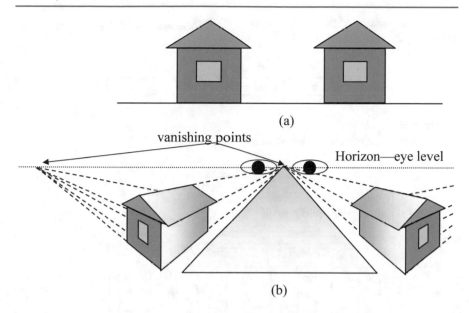

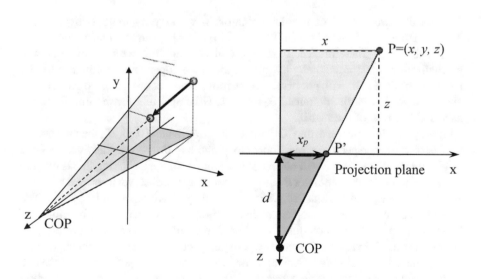

Fig. 3.26 Parallel orthographic and central projections of the same 3D scene

Fig. 3.27 Perspective projection of a point

point P with coordinates (x, y, z) is being projected onto point P' on plane $z = 0$ (axis Y is pointing up) from the *center of projection* (COP) located on a positive part of axis Z at distance d from the origin.

The coordinates of the projected point P' are $(x_p, y_p, 0)$. By considering similar triangles, we can obtain coordinate x_p as follows:

$$\frac{x_P}{d} = \frac{x}{d-z} \Rightarrow x_P = \frac{xd}{d-z} \Rightarrow x_P = \frac{x}{1-\frac{z}{d}} = \frac{x}{1+zr} \tag{3.56}$$

where $r = -1/d$.

Similarly, coordinate y_p can be derived:

$$y_P = \frac{y}{1+zr} \tag{3.57}$$

Putting it all together into a matrix form yields

$$x_P = \frac{x}{1+zr}, \ y_P = \frac{y}{1+zr}, \ r = -1/d$$

$$\begin{bmatrix} x' \\ y' \\ 0 \\ 1 \end{bmatrix} = \begin{bmatrix} \frac{x}{1+zr} \\ \frac{y}{1+zr} \\ 0 \\ 1 \end{bmatrix} = \begin{bmatrix} x \\ y \\ 0 \\ 1+rz \end{bmatrix} = \begin{bmatrix} 1 & 0 & 0 & 0 \\ 0 & 1 & 0 & 0 \\ 0 & 0 & 0 & 0 \\ 0 & 0 & r & 1 \end{bmatrix} \begin{bmatrix} x \\ y \\ z \\ 1 \end{bmatrix} \tag{3.58}$$

Note that we got rid of the denominator in x and y by multiplying all four homogeneous coordinates by $rz + 1$ that eventually changed the fourth coordinate from 1 to $rz + 1$. Therefore, each time the matrix is applied, we have to perform a normalization transformation by dividing x and y homogeneous coordinates by the fourth coordinate. It will preserve our homogeneous coordinates convention to always keep the fourth coordinate equal to 1. Note also that for the first time, we used the last row of the matrix.

If we place the observer at the COP point with coordinates $(0, 0, d)$, then the matrix will perform a perspective projection onto plane $z = 0$. All the lines which were originally parallel to axis Z will converge at point $(0, 0, -d)$. Proving this is outside the scope of this book. The larger d is, the farther away the observer is located from the projection plane, and the smaller scaling of x and y coordinates will be done. The smaller d is, the closer to the projection plane the observer is located, and therefore, the larger changes to x and y coordinates will be done. This provides the opportunity of creating bizarre 3D views of the scenes where perspective is outlandishly exaggerated.

For the arbitrary located COP and the projection plane, we can combine the matrix of perspective projection transformation onto $z = 0$ with a few translation and rotation transformations which will eventually bring the COP onto axis Z and the projection plane onto plane $z = 0$ (Fig. 3.28).

You might have wondered what could happen if the other two elements in the last row of the matrix will be used instead of being set to zero. Mathematically, it will make two or even three COPs with the respective vanishing points. It corresponds to so-called *two-* and *three-point perspective transformations*.

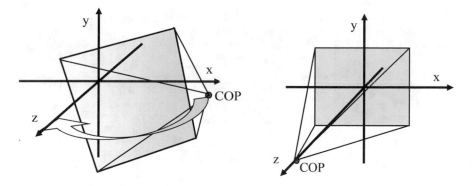

Fig. 3.28 Arbitrary perspective projection

Two-point perspective transformation:

$$\begin{bmatrix} x' \\ y' \\ 0 \\ 1 \end{bmatrix} = \begin{bmatrix} 1 & 0 & 0 & 0 \\ 0 & 1 & 0 & 0 \\ 0 & 0 & 0 & 0 \\ 0 & q & r & 1 \end{bmatrix} \begin{bmatrix} x \\ y \\ z \\ 1 \end{bmatrix} = \begin{bmatrix} x \\ y \\ 0 \\ qy + rz + 1 \end{bmatrix} \equiv \begin{bmatrix} \frac{x}{qy + rz + 1} \\ \frac{y}{qy + rz + 1} \\ 0 \\ 1 \end{bmatrix} \quad (3.59)$$

Three-point perspective transformation:

$$\begin{bmatrix} x' \\ y' \\ 0 \\ 1 \end{bmatrix} = \begin{bmatrix} 1 & 0 & 0 & 0 \\ 0 & 1 & 0 & 0 \\ 0 & 0 & 0 & 0 \\ p & q & r & 1 \end{bmatrix} \begin{bmatrix} x \\ y \\ z \\ 1 \end{bmatrix}$$

$$= \begin{bmatrix} x \\ y \\ 0 \\ px + qy + rz + 1 \end{bmatrix} \equiv \begin{bmatrix} \frac{x}{px + qy + rz + 1} \\ \frac{y}{px + qy + rz + 1} \\ 0 \\ 1 \end{bmatrix} \quad (3.60)$$

The respective centers of projection are located on the X-axis at $(-1/p, 0, 0)$, on the Y-axis at $(0, -1/q, 0)$, and on the Z-axis at $(0, 0, -1/r)$. Three vanishing points are located on the X-axis at $(1/p, 0, 0)$, on the Y-axis at $(0, 1/q, 0)$, and on the Z-axis at $(0, 0, 1/r)$. The examples of these projections as well as the previously considered single-point perspective projection are displayed in Fig. 3.29.

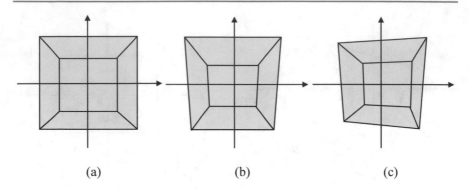

Fig. 3.29 **a** Single-, **b** two-, and **c** three-point perspective transformations of a cube

3.4.6   Axonometric Perspective Projections

Though mathematically correct, it is in practice difficult to manipulate two or three parameters to achieve the desirable appearance of the projection. Normally, axonometric rotations followed by a single-point perspective projection are used to achieve the same appearance as that of two- or three-point perspective projections (Fig. 3.30).

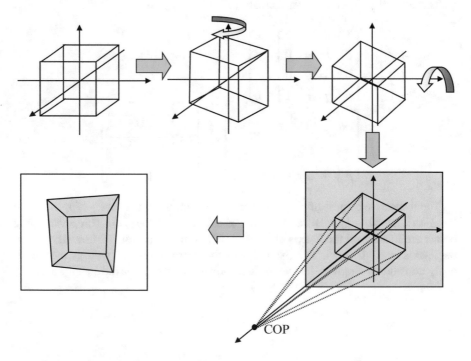

Fig. 3.30 Axonometric rotations followed by a single-point perspective projection

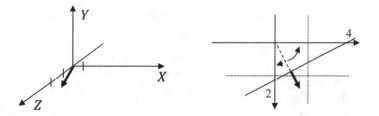

Fig. 3.31 Projection onto a plane defined by equation $x + 2z - 4 = 0$.

3.4.7 Projections on Any Plane

Let's consider an example of deriving a single matrix for a parallel orthographic projection onto a plane defined by the equation $x + 2z - 4 = 0$. The normal vector to this plane is $[1\ 0\ 2]$ (Fig. 3.31).

Since the location of the origin of the projected coordinate system is not specified for us, it can be any point on the projection plane. Therefore, the shortest way to solve the problem is to rotate the plane towards either the XY or YZ coordinate plane and then perform the respective projection onto this plane.

There can be two possible solutions. One is in x and y coordinates and another one for y and z coordinates:

$$
\begin{bmatrix} 1 & 0 & 0 & 0 \\ 0 & 1 & 0 & 0 \\ 0 & 0 & 0 & 0 \\ 0 & 0 & 0 & 1 \end{bmatrix}
\begin{bmatrix} \frac{2}{\sqrt{5}} & 0 & -\frac{1}{\sqrt{5}} & 0 \\ 0 & 1 & 0 & 1 \\ \frac{1}{\sqrt{5}} & 0 & \frac{2}{\sqrt{5}} & 0 \\ 0 & 0 & 0 & 1 \end{bmatrix}
=
\begin{bmatrix} \frac{2}{\sqrt{5}} & 0 & -\frac{1}{\sqrt{5}} & 0 \\ 0 & 1 & 0 & 1 \\ 0 & 0 & 0 & 0 \\ 0 & 0 & 0 & 1 \end{bmatrix}
\tag{3.61}
$$

$$
\begin{bmatrix} 0 & 0 & 0 & 0 \\ 0 & 1 & 0 & 0 \\ 0 & 0 & 1 & 0 \\ 0 & 0 & 0 & 1 \end{bmatrix}
\begin{bmatrix} \frac{1}{\sqrt{5}} & 0 & \frac{2}{\sqrt{5}} & 0 \\ 0 & 1 & 0 & 1 \\ -\frac{2}{\sqrt{5}} & 0 & \frac{1}{\sqrt{5}} & 0 \\ 0 & 0 & 0 & 1 \end{bmatrix}
=
\begin{bmatrix} 0 & 0 & 0 & 0 \\ 0 & 1 & 0 & 1 \\ -\frac{2}{\sqrt{5}} & 0 & \frac{1}{\sqrt{5}} & 0 \\ 0 & 0 & 0 & 1 \end{bmatrix}
\tag{3.62}
$$

Let's make one more exercise to implement in a matrix form the perspective projection from the viewer's position at $(1, 1, 3)$ onto the projection plane defined by the following three points $(2, 0, 0)$, $(0, 3, 1)$, $(2, 3, 0)$. With reference to Fig. 3.32, we may first translate the scene by $0, -1, -1$ so that the projection plane will pass through the origin and the observer will be placed on plane $y = 0$.

Then, we will rotate by angle $-\alpha$ so that the plane will become equal to $z = 0$ and the observer will be placed on axis Z. Finally, we will project onto plane $z = 0$ from $\text{COP} = \left(0, 0, \sqrt{2^2 + 1^2}\right) = \left(0, 0, \sqrt{5}\right)$.

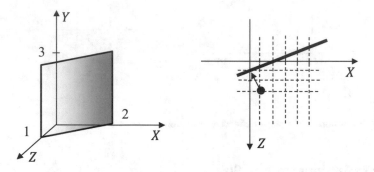

Fig. 3.32 Perspective projection onto a plane defined by three points

$$\cos\alpha = \frac{2}{\sqrt{2^2+1^2}} = \frac{2}{\sqrt{5}} \qquad \sin\alpha = \frac{1}{\sqrt{2^2+1^2}} = \frac{1}{\sqrt{5}}$$

$$\begin{bmatrix} 0 & 0 & 0 & 0 \\ 0 & 1 & 0 & 0 \\ 0 & 0 & 1 & 0 \\ 0 & -1 & -1 & 1 \end{bmatrix} \begin{bmatrix} \frac{2}{\sqrt{5}} & 0 & \frac{1}{\sqrt{5}} & 0 \\ 0 & 1 & 0 & 1 \\ -\frac{1}{\sqrt{5}} & 0 & \frac{2}{\sqrt{5}} & 0 \\ 0 & 0 & 0 & 1 \end{bmatrix} \begin{bmatrix} 1 & 0 & 0 & 0 \\ 0 & 1 & 0 & 1 \\ 0 & 0 & 0 & -\frac{1}{\sqrt{5}} \\ 0 & 0 & 0 & 1 \end{bmatrix} \qquad (3.63)$$

3.4.8 Viewing Frustum

Like in 2D, the scene can be larger than its portion which is to be visualized. It especially applies to large virtual environments where only a little portion of a scene is visualized at each time. To be able to cut away those parts of the scene that should not be visualized, a so-called *Viewing Volume* is used. It bounds the portion of the scene that is to be projected onto the view plane. The viewing volume is defined by front and back clipping planes. It is also often called *Viewing Frustum* or *Viewing Pyramid* (Fig. 3.33). The field of view can be controlled by fixing limits for the *x*- and *y*-coordinates visible to the observer.

3.4.9 Stereo Projection

To increase the perception of three-dimensional depth in a scene, a *stereo projection* can be used. It produces an image with characteristics analogous to those for true binocular vision. All the available stereography techniques depend upon supplying the left and right eyes with separate images. It requires projection of an object onto a plane from two different COPs, one for the right eye and another for the left eye. It can be achieved by translating the object to the left and to the right followed by projecting it (Fig. 3.34).

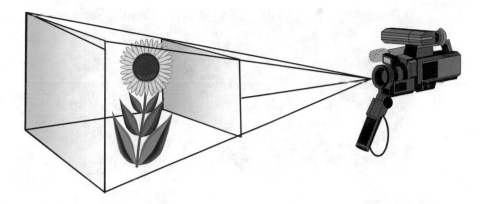

Fig. 3.33 Viewing volume

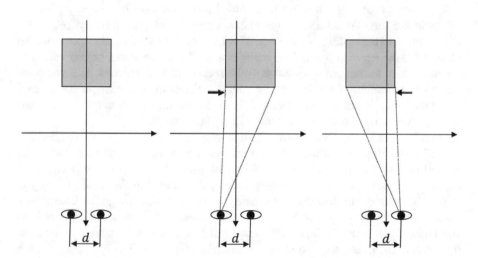

Fig. 3.34 Simple stereo-pair viewing method

In a matrix form, the stereo projection will require two different translations composed with projection transformations:

Left eye image:
$\begin{bmatrix} 1 & 0 & 0 & d/2 \\ 0 & 1 & 0 & 0 \\ 0 & 0 & 0 & 0 \\ 0 & 0 & r & 1 \end{bmatrix}$
Right eye image:
$\begin{bmatrix} 1 & 0 & 0 & -d/2 \\ 0 & 1 & 0 & 0 \\ 0 & 0 & 0 & 0 \\ 0 & 0 & r & 1 \end{bmatrix}$

$$(3.64)$$

where d is the distance between the COPs.

Fig. 3.35 Stereo pair

Distance d is actually an equivalent of the human *inter-ocular distance* which is of about 6.5 cm. To achieve a reasonable stereo effect, this distance has to be consistent with the scaling of the scene. The amount of 1/30 of the distance from the observer to the nearest object of the scene is a good average value. Larger values of d will produce a so-called *hyper-stereo*—the scene will look reduced, as if you were a giant. A reduction of d will cause an enlargement of the scene (*hypo-stereo*) that will give you the sensation of being a fly. It is recommended not to let the two camera axes converge at a point, but to keep them parallel.

You have to render the images with a resolution of 200 × 200 pixels. Leave a strip of about 10 pixels between the two images. The complete stereo pair (with 410 × 200 pixels) will therefore have a width of about 13 cm. So, it can easily be captured by the eyes. An example of a stereo pair is given in Fig. 3.35. When you look at the stereo-pair, just defocus your eyes. Try to look through the images at a point behind the upper ball like if you were daydreaming. You will then be able to see a third ghost image in the middle between the two images. You have to focus on this central image and a stereo image will magically appear! The left image for the left eye and the right image for the right eye just merge into a single 3D image. You may also need to move the image closer or further away from you to facilitate merging the images. Do not get upset if it does not work—you simply never used your eye muscles that way before. An alternative way is to use 3D glasses [5].

3.5 Summary

- Affine transformations are linear transformations of coordinates. They preserve similarity of the objects—the parallel lines remain parallel after transformations while sizes and angles are not preserved. There are only three basic affine transformations: translation, rotation, and scaling. All other linear transformations can be derived as a composition of these affine transformations.

- The matrix form of representation is chosen because multiple transformations can be combined in one single matrix that saves memory and time when big number of points has to be transformed.
- Homogeneous coordinates are used to allow translation to be represented in a matrix form together with rotation and scaling transformations.
- 2D affine transformations require a 3×3 matrix, 3D—4×4 matrix in homogeneous form.
- We use homogeneous coordinate "1" because it allows us to avoid the normalization transformation each time when matrix transformations are applied.
- Rotation transformation in 2D can be performed about a point while rotation in 3D is performed about an axis. Reflection in 2D can be done about a point or a line while in 3D about a point, line, or plane.
- Great care must be taken when rotation and scaling transformations are used. The order of transformations is important, and different results will be produced if the order of transformations changes.
- To visualize shapes, we have to project them from 3 to 2D space. All the projection transformations constitute two large groups: parallel and central projections.
- To project onto a projection plane of arbitrary location, the same transformations as for the rotation about an arbitrary axis are made to bring the normal of the projection plane to one of the coordinate axes.
- Two- and three-point central projections can be replaced by a central axonometric projection where the shape rotates about its vertical axis, followed by the rotation about its horizontal axis, and thereafter by the one-point central projection onto one of the coordinate planes.

References

1. Ostermann, A. and Wanner, G., Geometry by Its History, Springer, 2012.
2. Stillwell, J, The Four Pillars of Geometry, Springer, 2005.
3. Rogers, D. and Adams, J.A., Mathematical Elements for Computer Graphics, 2nd ed., McGraw-Hill, 1989.
4. Richter-Gebert, J., Perspectives on Projective Geometry, Springer, 2011.
5. How 3D PC Glasses Work, https://computer.howstuffworks.com/3d-pc-glasses.htm.

Motions

4

4.1 Animating Geometry

Moving geometric objects or motions are subject of *Computer Animation*. Generally, the term *Computer Animation* refers to creating moving images via the use of computers [1–3]. These moving images can show changes in a shape's position and size, as well as in its color, transparency and texture. The shapes can also perform *morphing*—transformation of one shape into another. Lighting condition or a camera position in a scene can also change in time to produce computer animation. Computer animation can be created and displayed in real time at a rate compatible with the refresh rate, as well as frame-by-frame, where each image is separately generated, stored, and then displayed one by one. In either case, the illusion of motion is eventually achieved by a rapid succession of individual images which are merged into a continuous motion by our brain as it was discussed in Chap. 1.

In Chaps. 2 and 3, we learnt how to define and transform objects. The parametric definitions of curves can be converted to definitions of moving points. Definitions of surfaces and solids can become time dependent. The affine transformations may be considered as not instantaneous changes of coordinates but performed through time. With all these methods we will be able to define moving or changing objects.

4.1.1 Motion of Points

Let's consider the motion of individual points.

Parametric representations of curves used in Chap. 2 define curves as sets of points created by sampling the modeling space with a point moving from the starting to finishing position:

$$x = f_1(u), \quad y = f_2(u), \quad z = f_3(u), \quad u \in [0, 1] \tag{4.1}$$

© The Author(s), under exclusive license to Springer Nature Switzerland AG 2021
A. Sourin, *Making Images with Mathematics*, Undergraduate Topics
in Computer Science, https://doi.org/10.1007/978-3-030-69835-5_4

When drawing curves, we increment the value of parameter u and generate vertices which have to be joined by segments. These definitions can be easily converted to definitions of coordinates of a moving point if the parameter u is linked to the value of time which is used in the modeling system:

$$u = f(t), \quad t \in [t_1, t_2], \quad u \in [0, 1] \tag{4.2}$$

Then, for each value of time t we will have a value of parameter u that will yield the coordinates of the moving point at the time t. The Eq. (4.2) can be then rewritten as

$$u(t) = \tau, \quad \tau \in [0, 1], \quad \tau = f(t), \quad t \in [t_1, t_2] \tag{4.3}$$

For the motion with a *uniform speed*, the time function $\tau = f(t)$ which is mapping the time t to the parameter u has to be linear, while for the motions with *acceleration* and *deceleration*—non-linear (Fig. 4.1).

The time is usually sampled with regular intervals (e.g. frame by frame). Then, the values of the parameter will be also equally spaced for the uniform speed, however sampled with increasing distance for acceleration, and with decreasing distance for deceleration. For the uniform speed function (Fig. 4.1a), a simple linear interpolation can be used:

$$\tau = (t - t_1)/(t_2 - t_1), \quad t \in [t_1, t_2] \tag{4.4}$$

For the precise definition of acceleration function (Fig. 4.1b), we have to derive it using the formulas of Newtonian physics of motion, as it will be later considered. To make a *pseudo-physical* definition, which defines motion that is like real but is based on simpler non-physical assumptions, we can use any function with a similar behavior, for example, a quadric function:

$$\tau = ((t - t_1)/(t_2 - t_1))^2, \quad t \in [t_1, t_2] \tag{4.5}$$

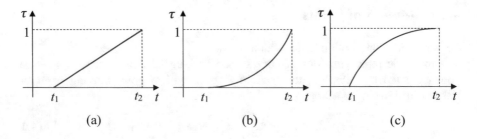

Fig. 4.1 Defining time mapping functions $\tau = f(t)$ for motion with **a** a uniform speed, **b** acceleration, and **c** deceleration

or a trigonometric function:

$$\tau = 1 - \cos(0.5\pi(t - t_1)/(t_2 - t_1)), \quad t \in [t_1, t_2] \tag{4.6}$$

Similarly, the deceleration functions can be defined by a quadric function:

$$\tau = 1 - ((t - t_1)/(t_2 - t_1))^2, \quad t \in [t_1, t_2] \tag{4.7}$$

or by a trigonometric function:

$$\tau = \sin(0.5\pi(t - t_1)/(t_2 - t_1)), \quad t \in [t_1, t_2] \tag{4.8}$$

Thus, when the straight line segment parametric equations are sampled with a uniform speed function, acceleration, and deceleration the points will be sampled as in Fig. 4.2.

Back and forth motion or *swing animation* (Fig. 4.3a) with a uniform speed can be defined by

$$\tau = 1 - |2(t - t_1)/(t_2 - t_1) - 1|, \quad t \in [t_1, t_2] \tag{4.9}$$

or by a trigonometric function which defines deceleration followed by acceleration (Fig. 4.3b):

$$\tau = \sin(\pi(t - t_1)/(t_2 - t_1)), \quad t \in [t_1, t_2] \tag{4.10}$$

This simple method of defining a uniform and non-uniform speed of motion is, however, lacking a mechanism of controlling the behavior. For instance, it is not easy to make the acceleration part of the function different from the deceleration portion, because symmetry is the very essence of the trigonometric functions. However, it has many applications to define motions of rigid bodies such as in the

Fig. 4.2 Sampling a point motion along a straight line with **a** a uniform speed, **b** acceleration and **c** deceleration

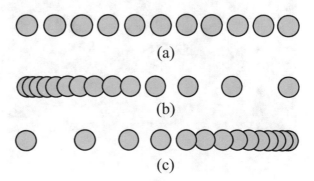

(a)

(b)

(c)

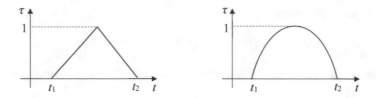

Fig. 4.3 Defining swing animation for motion with **a** a uniform speed, **b** deceleration and acceleration

example of a bouncing ball illustrated in Fig. 4.4. Here, the ball is defined as a shiny red parametric sphere with radius r bounces 12 times on a ground plane while linearly reducing the height of bouncing till it comes to rest. The ball is followed by a shadow cast from the directional light source placed above the ground plane.

To define the ball motion, we will use an absolute value of trigonometric *sin* function and will modulate it by a linear function. The coordinates of the center of the ball (a parametric sphere with radius r) are then defined by

$$
\begin{aligned}
x(t) &= maxdistance * \tau \\
y(t) &= maxheight * (1 - \tau) * abs(\sin(12\pi\tau)) + r \\
z(t) &= 0 \\
\tau &= (t - t_1)/(t_2 - t_1), \quad t = [0, 1]
\end{aligned}
\tag{4.11}
$$

The shadow will be defined as a plane disk following the ball on the ground plane (Fig. 4.5). Its radius r_d and variable color will change depending on the height of the ball. The variable color index will map to the $r\ g\ b$ color values by linear interpolation between the three pattern key values.

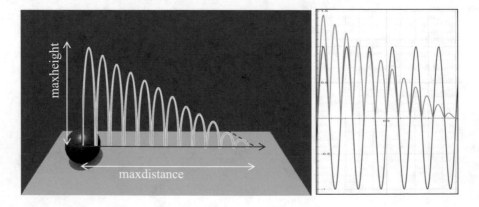

Fig. 4.4 Bouncing ball

$$r_d(t) = r_0 + r_0(1 - \tau) \cdot |\sin(12\pi\tau)|$$

$$shadow\_disk\_displacement(t) = maxdistance \cdot \tau$$

$$color\ index = \sqrt{(x - maxdist \cdot \tau)^2 + z^2} \qquad (4.12)$$

$$pattern\ key = [0.0 \quad 0.4 \quad 1.0]$$

$$pattern\ color = [0\ 0\ 0 \quad 0.2\ 0.8\ 0.2 \quad 0\ 1\ 0]$$

4.1.2 Animating Shape Definitions

The time variable can be plugged into any part of shape definition to turn the shapes into moving objects or changing their geometry.

Thus, the cylinder (Fig. 4.6) defined by

$$x = 2\sin(2\pi u)$$
$$y = 2\,v\tau$$
$$z = 2\cos(2\pi u) \qquad (4.13)$$
$$\tau = 1 - |2(t - t_1)/(t_2 - t_1) - 1|$$
$$u, v \in [0, 1], \quad t \in [t_1, t_2]$$

will change its height from 0 to 2 and then back to 0 from time t_1 to t_2.

Rotational sweeping of a sphere discussed in Chap. 2, can be illustrated by animating one of the rotation angles as shown in Fig. 4.7.

Fig. 4.5 Simulated shadow

Fig. 4.6 Growing cylinder

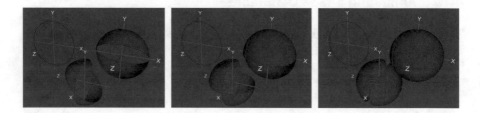

Fig. 4.7 Animating rotational sweeping of a sphere

$$
\begin{aligned}
x &= 3.0\cos(2\pi u)\\
y &= 3.0\sin(2\pi u)\cos(\pi v\tau)\\
z &= 3.0\sin(2\pi u)\sin(\pi v\tau)\\
\tau &= (t-t_1)/(t_2-t_1)\\
u,v &\in [0,1],\quad t\in[t_1,t_2]
\end{aligned}
\tag{4.14}
$$

4.1.3   Time-Dependent Affine Transformations

Object motions can be also defined with time-dependent affine transformations:

$$
\begin{bmatrix} x(t)\\ y(t)\\ z(t)\\ 1 \end{bmatrix}
=
\begin{bmatrix}
a(t) & b(t) & c(t) & l(t)\\
d(t) & e(t) & f(t) & m(t)\\
g(t) & h(t) & k(t) & n(t)\\
0 & 0 & 0 & 1
\end{bmatrix}
\begin{bmatrix} x\\ y\\ z\\ 1 \end{bmatrix}
\tag{4.15}
$$

Here, each element of the general affine transformation matrix is a function of time. The same time mapping function $\tau(t)$ can be applied to define motion with a uniform speed, acceleration, and deceleration.

Thus, translational motion can be defined with the matrix:

$$\begin{bmatrix} x(\tau) \\ y(\tau) \\ z(\tau) \\ 1 \end{bmatrix} = \begin{bmatrix} 1 & 0 & 0 & L\tau \\ 0 & 1 & 0 & M\tau \\ 0 & 0 & 1 & N\tau \\ 0 & 0 & 0 & 1 \end{bmatrix} \begin{bmatrix} x \\ y \\ z \\ 1 \end{bmatrix} \quad \tau \in [0,1] \tag{4.16}$$

where L, M, and N are the values of the final displacement of x, y, and z coordinates. Rotation about axis X by angle α will be defined:

$$\begin{bmatrix} x(\tau) \\ y(\tau) \\ z(\tau) \\ 1 \end{bmatrix} = \begin{bmatrix} 1 & 0 & 0 & 0 \\ 0 & \cos\alpha\tau & -\sin\alpha\tau & 0 \\ 0 & \sin\alpha\tau & \cos\alpha\tau & 0 \\ 0 & 0 & 0 & 1 \end{bmatrix} \begin{bmatrix} x \\ y \\ z \\ 1 \end{bmatrix} \quad \tau \in [0,1] \tag{4.17}$$

Scaling by A, B, and C will be implemented with

$$\begin{bmatrix} x(\tau) \\ y(\tau) \\ z(\tau) \\ 1 \end{bmatrix} = \begin{bmatrix} 1+(A-1)\tau & 0 & 0 & 0 \\ 0 & 1+(B-1)\tau & 0 & 0 \\ 0 & 0 & 1+(C-1)\tau & 0 \\ 0 & 0 & 0 & 1 \end{bmatrix} \begin{bmatrix} x \\ y \\ z \\ 1 \end{bmatrix}, \quad \tau \in [0,1]$$

$$\tag{4.18}$$

Note, that in the scaling motion, the scaling coefficients linearly change from 1 to A, B, and C, respectively.

Similarly to how it was done with the ordinary affine transformation, several time-dependent transformations can be applied to an object to define its motion.

$$P\prime(t) = \mathbf{T}_n(t)\mathbf{T}_{n-1}(t)\ldots\mathbf{T}_2(t)\mathbf{T}_1(t)\mathbf{P}$$
$$\mathbf{T}_1(t), [t_1,t_2,], \ \mathbf{T}_2(t), [t_1,t_3,], \ \mathbf{T}_3(t), [t_2,t_3,], \ \mathbf{T}_4(t), [t_3,t_4,] \tag{4.19}$$
$$[t_1,t_2,] : \mathbf{T}_2\mathbf{T}_1, [t_2,t_3,] : \mathbf{T}_3\mathbf{T}_2, [t_3,t_4,] : \mathbf{T}_4$$

For each value of time t the respective matrix parameter will be evaluated, substituted into the matrix, and after that the combined matrix will be calculated. This matrix will transform the object to the location, size, and orientation corresponding to the value of t.

This method can be used for defining the motion of rigid objects. In that case, each location in space will require us to rotate the basic model of the object about three axes followed by its translation to the respective point in 3D space. For example, a flying airplane can be modeled this way (Fig. 4.8). These rotations—yaw, pitch, and roll—will be discussed in the Chap. 6.

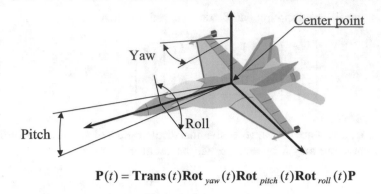

$$\mathbf{P}(t) = \mathbf{Trans}\,(t)\mathbf{Rot}_{yaw}\,(t)\mathbf{Rot}_{pitch}\,(t)\mathbf{Rot}_{roll}\,(t)\mathbf{P}$$

Fig. 4.8 Modeling a flying airplane

4.1.4 Shape Morphing

Besides using matrices, we can perform animated transformations between different shapes (*shape morphing*). When morphing from one shape to another, we have to interpolate vertices of the animated shapes. An example of such morphing for two shapes defined parametrically is given in Fig. 4.9.

Linear interpolation can be used for morphing implicitly defined shapes as well. In that case, for any two shapes G_1 and G_2 defined with implicit functions f_1 and f_2, the time-dependent morphing function for object $G(t)$ will be defined as follows:

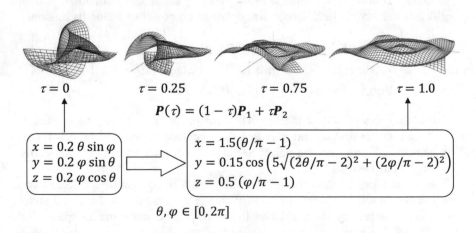

Fig. 4.9 Morphing with linear interpolation and parametric functions

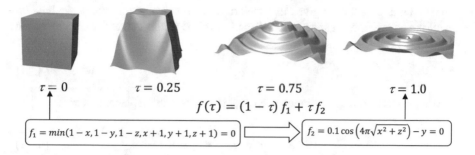

Fig. 4.10 Morphing with implicit functions

$$G_1: \quad f_1(x, y, z) = 0$$
$$G_2: \quad f_2(x, y, z) = 0$$
$$G(t): \quad f_1(x, y, z)(1 - \tau) + f_1(x, y, z)\tau = 0, \quad \tau \in [0, 1] \tag{4.20}$$

An example of morphing with implicit functions is given in Fig. 4.10.

4.2 Physically Based Motion Specifications

In the previous section we mostly relied upon visual impression to define various motions. Such technique cannot be used if we want better realism. In that case we must use physical laws rather than mimicking physical behaviors.

When considering realistic animation of objects, we have to take into consideration *Newton's Laws of Motion* [4]. Although these laws define the motion of particles that have mass but no size, they can be employed for describing the motion of large objects in computer animation [5].

Let's recall that the first Newton's law states that an object moves in a straight line, unless disturbed by some force. The second law states that a particle of mass m, subjected to a force F, moves with acceleration F/m. Finally, the third law is useful when considering collisions between objects, and it states that if a particle exerts a force on a second particle, the second particle exerts an equal force in the opposite direction.

Let us examine how these laws can help us with defining some simple motions.

4.2.1 Motion with Constant Acceleration

There are many computer animation applications, in which objects move with a constant acceleration. It is called *uniformly accelerated motion* or *motion with constant acceleration*. A common problem here is to determine the velocity of an object after a certain time given its acceleration. To solve this problem, we have to recall the equation for a speed v as a function of the initial velocity v_0, acceleration a, and time t.

$$v = v_0 + at \tag{4.21}$$

We may also use an average speed v_{av}, to define the displacement x:

$$v_{av} = \frac{v_0 + v_{final}}{2}, \quad x = v_{av}t, \quad x = x_0 + \frac{(v_0 + v_{final})t}{2} \tag{4.22}$$

Substituting $v_{final} = v_0 + at$ into the equation for x yields

$$x = x_0 + v_0 t + \frac{at^2}{2} \tag{4.23}$$

where x_0 is the initial position. From here, we can derive an equation for speed v, as a function of the initial velocity v_0, acceleration a, the displacement x, and the initial location x_0.

$$t = \frac{v - v_0}{a} \quad x - x_0 = \frac{v^2 - v_0^2}{2a} \quad v^2 = v_0^2 + 2a(x - x_0) \tag{4.24}$$

With these equations, we can solve such problems as "how far the object moves before stopping", and "how long it takes to stop".

Let's consider an example of a ball moving by a horizontal surface. The motion begins with an initial velocity of 2 m/s and the ball slows down with a constant negative acceleration of $a = -0.2$ m/s^2.

The displacement function is $x(t) = x_0 + 2t - 0.2t^2/2$. To compute how long it will take before the ball stops, we have to do the following math:

$$v = 0 \,\text{m/s} \quad v_0 = 2 \,\text{m/s} \quad a = -0.2 \,\text{m/s}^2 \quad x = x_0 + \frac{v^2 - v_0^2}{2a}$$

$$x = x_0 + \frac{v^2 - v_0^2}{2a} = 0 + \frac{0^2 - 2^2}{2(-0.2)} = 10 \,\text{m}$$

$$t = \frac{v - v_0}{a} = \frac{0 - 2}{-0.2} = 10 \,\text{s}.$$

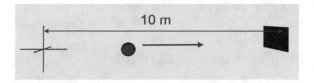

Fig. 4.11 Rolling ball

The motion then can be defined as a horizontal displacement of a sphere provided there are no any textures mapped onto its surface that would reveal that the sphere is not really rolling but only displacing (see Fig. 4.11).

$$x = 0.3 \cos u \cos v + 2\tau + \frac{-0.2\tau^2}{2}$$

$$y = 0.3 \cos u \sin v$$

$$z = 0.3 \sin u$$

$$\tau = \frac{t}{10}, \quad t \in [0, 10]$$

$$u \in [0, 2\pi], \quad v \in [0, \pi]$$

In another example, we will model a runaway motion of a jet plane that makes a takeoff in two periods. It starts motion with an acceleration of 3.6 m/s^2 that lasts 5.0 s. Then the afterburner engines are turned up to full power to achieve an acceleration of 5.1 m/s^2. Knowing that the speed needed for takeoff is 84.4 m/s, we have to calculate the length of a runway needed for it and the total time of takeoff. It involves the following math:

The 1st period : $\quad v = 0 + 3.6 \cdot 5.0 = 18$ m/s, $\quad x = 0 + 3.6 \cdot 5^2/2 = 45$ m

The 2nd period : $\quad t = \frac{v - v_0}{a} + t_0 = \frac{84.4 - 18}{5.1} = 18$ s

$$x = \frac{a(t - t_0)^2}{2} + v_0(t - t_0) + x_0 = \frac{5.1(18 - 5)^2}{2} + 18(18 - 5) + 45 = 710 \, \text{m}$$

The jet motion then can be modelled as follows:

$distance(t)$:

$if(t < = 5)\ return\ (3.6 \cdot t^2)/\ 2;$

$if(t > 5)\ return\ 45 + 18(t - 5) + \dfrac{5.1(t - 5)^2}{2} ; \}$

$height(t)$:

$\{if(t < = 18)\ return\ 0;$

$if(t > 18)\ return\ (t - 18)20; \}$

$t > 0$

4.2.2 Motion Under Gravity

Objects placed above the ground level are attracted back through a *gravitational force mg*, where m is the objects' mass, and g is the acceleration due to gravity. This force is called the *object's weight*. If the object is released with an initial velocity v_0, the distance d traveled after time t is defined by

$$d = v_0 t + \frac{gt^2}{2} \qquad (4.25)$$

Given the initial height of the object above the ground level, the height of the falling object will be calculated by

$$h = h_0 - v_0 t - \frac{gt^2}{2} \qquad (4.26)$$

One of the typical computer animation problems is to calculate time when the object collides with the plane defining the ground level. Let's consider a falling ball (Fig. 4.12a). If we are able to sample the height at various time intervals, the point of collision with the ground is detected when the ball's height is zero, or to be exact, when the height is equal to the ball's radius r. However, because of the discrete nature of the sampling process, it may become so that the current calculated ball's height h_c is less than r or even negative, while the previous value h_p is positive (Fig. 4.12b). The simplest solution in that case is to position the ball on the ground plane (Fig. 4.12c). This solution is feasible if the sampling rate was high compared with the ball's velocity. Alternatively, the exact impact time has to be computed by solving a quadratic equation (Fig. 4.12d).

$$v_c = v_p + gt, \quad h_p - v_p t - \frac{gt^2}{2} = r$$

$$t = \frac{v_p \pm \sqrt{v_p^2 + 2g(h_p - r)}}{-g} \Rightarrow t_{impact = tp + t}$$

$$(4.27)$$

where only the positive root is to be used.

Let's consider an example of a falling ball from a height of 100 m. The ball will reach the ground for 4.52 s according to:

$$Height = 100 - \frac{9.8t^2}{2}$$

$$0 = 100 - \frac{9.8t^2}{2} \Rightarrow t = 4.52 \, \text{s}$$

For a ball with radius 2, it then can be modeled as:

$$x = 2 \cos u\pi \, \cos v\pi$$

$$y = 2 \sin u\pi + 2 + 100 - \frac{9.8t^2}{2}$$

$$2 \cos u\pi \sin v$$

$$u, v \in [0, 1]$$

$$t \in [0, 4.52]$$

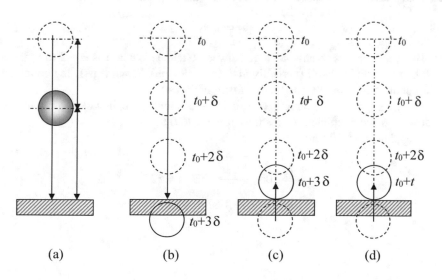

(a)   (b)   (c)   (d)

Fig. 4.12 Sampling of a falling ball

If the ball is first thrown up with a speed of 10 m/s from a height of 15 m, the equations will change as follows:

$$v = 0 = 10 - 9.8t \Rightarrow t = 1.02 \, \text{s}$$

$$h = h_0 + v_0 t - \frac{gt^2}{2} = 15 + 10 \cdot 1.02 - 9.8 \frac{1.02^2}{2} = 20.1 \, \text{m}$$

$$Height : \left\{ h = 15 + 10t - \frac{9.8t^2}{2}; \quad if(h < 0)h = 0; \quad return \, h; \right\}$$

$$t > 0$$

If the falling object is supposed to bounce back, its speed has to reverse, and the process will continue. Thus, in the case of perfectly elastic collision the bouncing of the ball dropped from 100 m in the previous example will continue indefinitely. It can be defined as:

$$n = floor\left(\frac{t}{4.52}\right); \quad i = n/2 - floor\left(\frac{n}{2}\right); \quad //making \, a \, switch$$

$$if \, (i == 0) \, \{d = 100; \, v_0 = 0;\} \, // \, going \, down$$

$$else \, \{d = 0; \, v_0 = 44.3;\} \, // \, going \, up$$

$$\tau = t - 4.52 \, n; \quad return \, d + v_0 \, \tau - \frac{9.8\tau^2}{2};$$

$$t > 0$$

In reality, this bouncing ball should come to rest sooner or later because of the conversion of its kinetic energy. To simulate this, a *coefficient of restitution k* is to be used which relates speed before and after the impact.

$$v_2 = k \, v_1 \tag{4.28}$$

This coefficient is to have a value between 0 and 1. When it is equal to 0, there will be no bouncing at all (perfectly *inelastic collision*). When it is 1, the bouncing will continue indefinitely (perfectly *elastic collision*).

Then, the previous example with indefinite bouncing ball will change to the following if the coefficient of restitution is set to 0.9:

Dropping from 100 m:

$$0 = 100 - \frac{9.8t^2}{2} \Rightarrow t = 4.52 \, \text{s}$$

$$v = 0 + 9.8 \cdot 4.52 = 44.3 \, \text{m/s}$$

Bouncing back with a reduced speed:

$$v_1 = 44.3 \cdot 0.9 = 39.87 \frac{m}{s} \quad v = 0 = 39.87 - 9.8t \quad t = \frac{39.87}{9.9} = 4.03 \; s$$

$$h = 39.87 \cdot 4.03 - 9.8 \cdot \frac{4.03^2}{2} = 81.1 \; m$$

Dropping from the reduced height of 81.1 m:

$$0 = 81.1 - \frac{9.8t^2}{2} \Rightarrow t = 4.07 \; s$$

$$v_2 = 0 + 9.8 \cdot 4.07 = 39.89 \; \text{m/s}$$

etc.

If the motion is not vertical, i.e. the object is thrown (projected) into the air, we consider a case of *projectiles*. The motion of projectiles is still going under the effect of gravity, and, therefore, there is only an acceleration due to gravity. For a 2D case (Fig. 4.13), it can be written as:

$$v_x = v_0 \cos \alpha + 0 \cdot t$$
$$v_y = v_0 \sin \alpha - gt$$
$$x = v_0 t \cos \alpha + \frac{0 \cdot t^2}{2}$$
$$y = v_0 t \sin \alpha - \frac{gt^2}{2}$$

In the ideal conditions without air resistance the object then will move by parabolic trajectory $y = ax^2 + bx + c$.

For example, if a ball is thrown into the air with 45° angle with the ground, its x and y coordinates will be computed as:

$$x = 15 \cdot t \cdot 0.7$$
$$y = 15 \cdot t \cdot 0.7 - \frac{9.8t^2}{2}; \quad if\,(y<0) \; y = 0;$$
$$t > 0$$

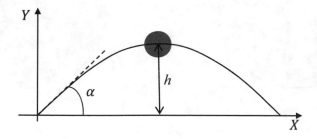

Fig. 4.13 Motion of a projectile

The total motion time T can be then computed taking into account that at time $t = \frac{T}{2}$ the vertical velocity is 0, hence:

$$0 = v_0 \sin \alpha - \frac{gT}{2} = > T = \frac{2v_0 \sin \alpha}{g}$$

The maximum height will be reached at time $t = \frac{T}{2}$:

$$h = v_0 t \sin \alpha - \frac{gt^2}{2} = \frac{(v_0 \sin \alpha)(v_0 \sin \alpha)}{g} - \frac{g\left(\frac{v_0 \sin \alpha}{g}\right)^2}{2} = \frac{v_0^2 (\sin \alpha)^2}{2g}$$

4.2.3 Rotation Motion

In Chap. 3, we discussed how rotational movements can be modeled using affine transformations. Let us now examine some of the physically-based methods of making rotating shapes in computer animation.

An object that moves in a circle at constant speed v is said to experience *uniform circular motion*. Speed v can be expressed through *angular speed* $w = d\theta/dt$ and radius r as it follows from Fig. 4.14a.

The circumference is $C = 2\pi r$. Then, any arc is $S = \theta r$. Therefore, the linear speed $v = dS/dt = rd\theta/dt$ which yields

$$v = wr \tag{4.29}$$

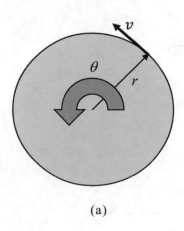

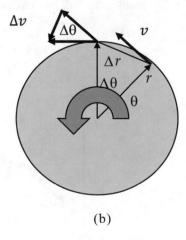

(a)   (b)

Fig. 4.14 Uniform circular motion

Although, the magnitude of velocity v remains constant, its direction is continuously changing as the object moves around the circle. Since acceleration is defined as the rate of change of velocity, a change in the direction of velocity constitutes an acceleration just as does a change in the magnitude of velocity. Thus, an object experiencing uniform circular motion is continuously accelerating. This acceleration is called *centripetal acceleration* or *center-seeking acceleration* since it is directed along the radius toward the center of the circle.

From Fig. 4.14b, it follows that

$$\frac{\Delta v}{v} = \frac{\Delta r}{r} \quad \frac{\Delta v}{v\Delta t} = \frac{\Delta r}{r\Delta t} \quad \frac{\Delta v}{\Delta t} = \frac{v\Delta r}{r\Delta t} \tag{4.30}$$

If $\Delta t 0$, then the center-seeking acceleration is $a = v\frac{v}{r} = \frac{v^2}{r}$ which yields

$$a = \frac{v^2}{r} = \frac{(wr)^2}{r} = rw^2 \tag{4.31}$$

Let's consider now the problem of animating a gears mesh. The smaller gear, rotating counter-clockwise, is a driving gear, while the other one is a driven gear and it rotates, respectively, clockwise (Fig. 4.15).

While the rotation of both gears can be defined with time-dependent rotation matrices, we only know the rotational speed w_1 of the driving gear, and thus know its angle of rotation $\alpha(t)$. We need to know the rotational speed of the second gear to be able to define its rotation angle function $\beta(t)$. At the point of contact P, the angular speeds of both gears are the same but have different directions. Therefore, the following relationship holds:

$$w_1 r_1 = -w_2 r_2 \tag{4.32}$$

Therefore,

$$w_2 = \frac{-w_1 r_1}{r_2} \tag{4.33}$$

Fig. 4.15 Rotating gears

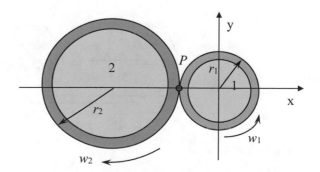

We can now define the angles of rotation for both gears and then substitute them into the relevant matrices:

$$\alpha(t) = 2\pi kt, \quad \beta(t) = -2\pi kt \frac{r_1}{r_2} \tag{4.34}$$

where the driving gear makes k rotations per second.

The first gear will be animated with:

$$\begin{bmatrix} x\prime \\ y\prime \\ 1 \end{bmatrix} = \begin{bmatrix} \cos(2\pi kt) & -\sin(2\pi kt) & 0 \\ \sin(2\pi kt) & \cos(2\pi kt) & 0 \\ 0 & 0 & 1 \end{bmatrix} \begin{bmatrix} x \\ y \\ 1 \end{bmatrix} \tag{4.35}$$

while the second, driven gear, with the following matrix:

$$\begin{bmatrix} x\prime \\ y\prime \\ 1 \end{bmatrix} = \begin{bmatrix} \cos\left(-2\pi kt\frac{r_1}{r_2}\right) & -\sin\left(-2\pi kt\frac{r_1}{r_2}\right) & -(r_1 + r_2) \\ \sin\left(-2\pi kt\frac{r_1}{r_2}\right) & \cos\left(-2\pi kt\frac{r_1}{r_2}\right) & 0 \\ 0 & 0 & 1 \end{bmatrix} \begin{bmatrix} x \\ y \\ 1 \end{bmatrix} \tag{4.36}$$

Now, let's consider a wheel rolling across a flat surface (Fig. 4.16). We can define the linear displacement of the wheel as a function of time $d(t) = vt$, where v is the linear speed. We need to derive the respective rotation angle $\alpha(t)$.

The relationship linking the linear and rotational speeds can be derived from the definition of angular speed, and from the observation that the distance d is equal to the arc which point P rotates on the circle.

$$d = \alpha r \quad \alpha(t) = \frac{d}{r} = \frac{vt}{r} \tag{4.37}$$

The matrix transformation for the wheel will be the following:

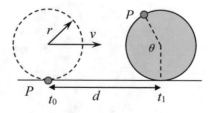

Fig. 4.16 Rolling wheel

$$\begin{bmatrix} x\prime \\ y\prime \\ 1 \end{bmatrix} = \begin{bmatrix} 1 & 0 & vt \\ 0 & 1 & 0 \\ 0 & 0 & 1 \end{bmatrix} \begin{bmatrix} \cos\frac{vt}{r} & -\sin\frac{vt}{r} & vt \\ \sin\frac{vt}{r} & \cos\frac{vt}{r} & 0 \\ 0 & 0 & 1 \end{bmatrix} \begin{bmatrix} x \\ y \\ 1 \end{bmatrix} \qquad (4.38)$$

4.3   Summary

- For motion by path of rigid objects, either motion by parametric curves or time-dependent affine transformation can be used.
- Function-based animation uses the same linear transformation applied to the parametric or implicit functions rather than individual vertices.
- Different time functions can simulate pseudo-physically motions with constant speed, acceleration, deceleration as well as swing (back and forth) motions.
- Physical laws are to be used when better realism of motion is to be achieved. Physical- and pseudo-physical approaches have their advantages and disadvantages, and the choice of methods depends on the application problem. Generally, the physically-based methods are computationally more expensive.

References

1. How Computer Animation Works, https://entertainment.howstuffworks.com/computer-animation.htm.
2. Thalmann, N.M., Thalmann, D.,Computer Animation. Theory and Practice, Springer, 1990.
3. Watt, A. and Watt, M., Advanced Animation and Rendering Techniques. Theory and Practice, Addison-Wesley, 1992.
4. Fishbane, P., Gasiorowicz, S., Thornton, S., Physics, 2nd ed., Prentice Hall, 1996.
5. Vince, J., Virtual Reality systems, Addison-Wesley, 1995.

Adding Visual Appearance to Geometry

5

5.1 Illumination

In the previous chapters, we have discussed how to define and transform geometric models of the shapes. However, this is not enough to make graphics images. We also have to define how exactly this geometry should look when it is rendered into graphics images. We have to define what colors, material properties, and textures have to be associated with each geometric object, and then determine what will be the respective values of the color assigned to each pixel of the graphics image. In order to do that, first of all, we have to define at least one light source in the scene. The task of *illumination* will be then to determine how the light emitted by the light sources reflects from surfaces of the shapes and produces what we perceive as a color. This task will include *lighting* and *shadingShading*. Lighting is a computation of the luminous intensity reflected from a specified 3D point. Shading is assigning colors to pixels.

There are several common types of light sources used in computer graphics. They are *ambient light, directional, point, spotlight*, and *area light sources*.

Ambient light is light that comes from no particular direction. It has no spatial or directional characteristics. The amount of ambient light incident to each object is a constant for all surfaces in the scene. The amount of ambient light that is reflected by an object is independent of the object's position or orientation. Surface properties often determine how much of this ambient light is reflected.

A *directional light* source is located at an infinitely remote point and therefore, all rays of light coming from it can be considered as parallel lines. Only the color of the light source and the direction from a surface to the light source are important when the lighting problem is being solved. With a directional light source, this direction is a constant for every surface (Fig. 5.1).

A *point light* source is an infinitely small light source with a finite location in the scene so that all the rays appear to come from a point equally in all directions (Fig. 5.2).

© The Author(s), under exclusive license to Springer Nature Switzerland AG 2021
A. Sourin, *Making Images with Mathematics*, Undergraduate Topics
in Computer Science, https://doi.org/10.1007/978-3-030-69835-5_5

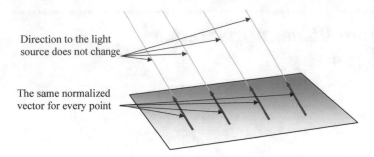

Fig. 5.1 Directional light source

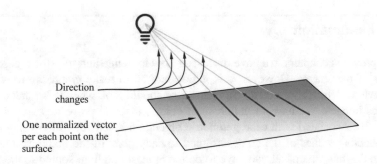

Fig. 5.2 Point light source

The rays emitted from a point light source radially diverge from it. The direction of the light to each point on the surface changes when the point light source is used. Thus, a normalized vector to the light emitter must be computed for each point that is illuminated. The location and the color of the light source are to be defined.

A *spotlight* is a point light source whose intensity falls off away from a given direction (Fig. 5.3). The location, color, direction, and parameters that control the rate of fall-off are to be defined for this light source.

An *area light* is a luminous 2D object which radiates light from all points on its surface (Fig. 5.4). Compared to the previously discussed light sources, it has a finite size, rather than be just a point in a 3D space.

5.2 Lighting

There can be two different ways to perform the lighting task. One is physically based. It assumes using models based on the actual physics of light's interactions with matter. An alternative method is using an *empirical illumination model* which

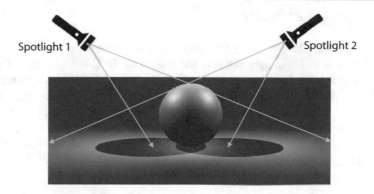

Fig. 5.3 Spotlight

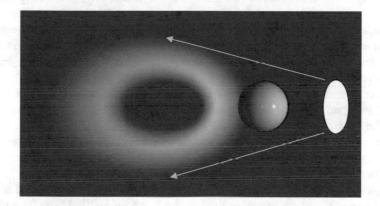

Fig. 5.4 Area light source

approximates the actual physics. We will consider the empirical illumination that provides quite acceptable results for most cases.

The empirical illumination model includes three components contributing to illumination calculated at any given point: *ambient*, *diffuse*, and *specular* reflections. Let's consider how these components can be calculated for any point $P = (x, y, z)$ on surface S. Our goal is to calculate a total illumination $I_p \in [0, 1]$ which will give us the amount of light reflected back from point P on the surface. This number then can be used to compute actual value of color visible at each point where illumination is calculated. If the color assigned to an object is r, g, b, the actual color at point P will be equal to it only where the illumination is maximum, i.e. 1. If the illumination is 0, the point will be assigned black color. For the values of illumination between 0 and 1, the color at point P will be computed as $I_p \cdot r$, $I_p \cdot g$, $I_p \cdot b$. If the calculated illumination is greater than 1, it has to be lowered down to 1.

5.2.1 Ambient Reflection

Ambient reflection results from the ambient light defined in the scene. If the intensity of the ambient light defined for the scene is $I_a \in [0, 1]$, then for each surface the amount of ambient reflection can be defined by an ambient reflection coefficient k_a so that the ambient contribution to the light reflected at point P will be computed as $I = k_a \cdot I_a$. If only ambient light is defined in the scene, we will not be able to see any shades but rather colored silhouettes of the shapes. An example of only ambient light used for displaying a sphere with different ambient reflection parameters k_a is given in Fig. 5.5.

5.2.2 Diffuse Reflection

Diffuse reflection from a surface assumes that the incident light is equally scattered in all directions independently of the viewer's position. The surface in that case is considered as an ideal diffuse reflector or *Lambertian reflector*. For such surfaces, the amount of the reflected incoming light energy is calculated with *Lambert's cosine law*. This law states that the reflected energy I_d from a small surface area in a particular direction is proportional to the cosine of the angle θ between that direction $\mathbf{L}$ and the surface normal $\mathbf{N}$. The I_L term defines the intensity of the incoming light. The k_d term represents the diffuse reflectivity of the surface (Fig. 5.6).

The amount of energy reflected in any direction is constant in this model. While it is independent of the viewing direction, it, however, depends on the light source's orientation relative to the surface.

In practice, we use vector analysis to compute the *cos* term indirectly. If both the normal vector $\mathbf{N}$ and the incoming light vector $\mathbf{L}$ are normalized (unit length), then the diffuse intensity can be computed as follows:

$$I_d = k_d I_L (\mathbf{N} \cdot \mathbf{L}) \tag{5.1}$$

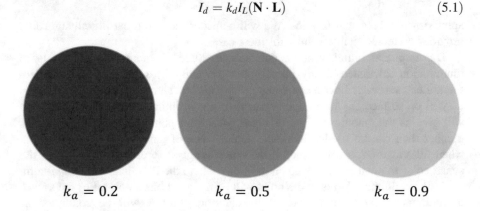

$$k_a = 0.2 \qquad\qquad k_a = 0.5 \qquad\qquad k_a = 0.9$$

Fig. 5.5 Ambient reflections with different ambient reflection parameters

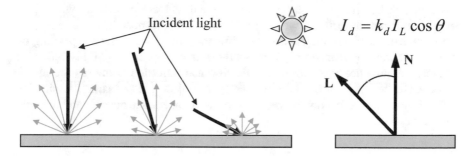

Incident light

$$I_d = k_d I_L \cos \theta$$

Fig. 5.6 Diffuse reflection

Thus, if $\mathbf{N} = [a\,b\,c]$ and $\mathbf{L} = [d\,e\,f]$, then $\mathbf{N} \cdot \mathbf{L} = \cos \theta = [a\,d \quad b\,e \quad c\,f]$.

We need only to consider angles from $0°$ to $90°$. Points with greater angles (where the dot product is negative) are blocked from the light source by the surface, and the respective reflected energy is 0.

By applying this formula to directional and point light sources, we can easily see that for the directional light sources the diffuse reflection will depend only on the normal to the surface, while for the point light sources it will depend both on the light vector and the normal.

The amount of energy reflected from different objects in a scene is controlled by the respective coefficients k_d that define the reflective property of the material. Examples of diffuse reflections with different parameters k_d are shown in Fig. 5.7.

5.2.3 Specular Reflection

If only the ambient and diffuse reflections are used in the scene, the shapes will have a rather dull appearance as if they were made of a material with a rough surface such as concrete or brick. If we needed to make a shiny surface, as if it were

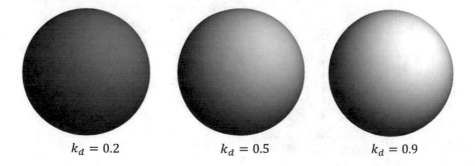

$k_d = 0.2$ $k_d = 0.5$ $k_d = 0.9$

Fig. 5.7 Diffuse reflections with different diffuse reflection parameters

made of polished metal or glass, the diffuse and ambient reflections will fail to simulate it. On such surfaces, we see highlights or bright spots. These bright spots are view-dependent, and they appear on different parts of the surface depending on where the surface is seen from. To make the surface appear as a highly reflective surface, we introduce *specular reflection*. Specular reflection is merely the mirror reflection of the light source in the surface of the object. This specular illumination component is to be a function of the observer location, surface normal, and the light source location.

5.2.4 Phong Illumination

One widely used function that approximates specular reflection is called the *Phong Illumination Model* (named after its inventor Phong Bui Tuong [1]). This model has no physical basis, yet it is one of the most commonly used illumination models in computer graphics. It assumes that the incoming ray, the surface normal, and the refracted ray all lie in a common plane. It also assumes that most of the reflected light travels in the direction of the ideal ray. However, because of microscopic surface variations, some of the light will be reflected with just a slight offset from the ideal reflected ray (Fig. 5.8). The cos term reaches the maximum value when the surface is viewed from the mirrored direction and falls off to 0 when viewed 90° away from it. The exponent n controls the rate of this fall-off. The effect of different values of n is illustrated in Figs. 5.9 and 5.10. The *cos* term of Phong's specular illumination can be found using the following relationship:

$$I_s = k_s I_L (\mathbf{V} \cdot \mathbf{R})^n \tag{5.2}$$

where $\mathbf{V}$ is the unit vector in the direction of the viewer and $\mathbf{R}$ is the mirror reflectance direction.

The vector $\mathbf{R}$ can be computed from the incoming light direction $\mathbf{L}$ and the surface normal $\mathbf{N}$ as it is illustrated in Fig. 5.11.

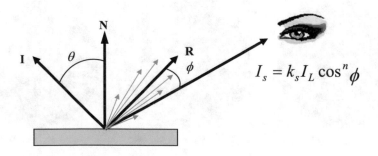

Fig. 5.8 Specular reflection

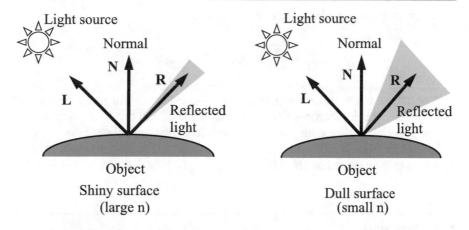

Fig. 5.9   Specular reflection of shiny and dull surfaces

$n = 50$ $n = 10$ $n = 2$

Fig. 5.10   Specular reflections of a sphere with different exponents n

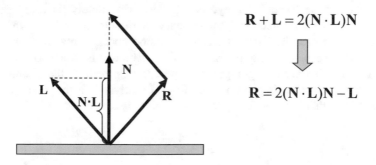

Fig. 5.11   Computing vector **R**

The final empirical illumination model then becomes

$$I = k_a I_a + \sum_{m}^{lights} I_{L_i}(k_d(\mathbf{N} \cdot \mathbf{L}) + k_s(\mathbf{V} \cdot \mathbf{R})^n) \tag{5.3}$$

The diffuse and specular components are to be calculated for every light source in the scene. This is a computationally expensive task, and this is why ray tracing, which is performed for each pixel, is a slow rendering technique. Faster computation can be done if Blinn–Phong reflection model [2] is used where a specular component is computed as $(\mathbf{N} \cdot \mathbf{H})^n$ where $\mathbf{H} = \frac{\mathbf{V}+\mathbf{L}}{\|\mathbf{V}+\mathbf{L}\|}$.

5.3   Shading

We have learned how to compute illumination at a point on a surface. But at which points on the surface is the illumination model to be applied? Illumination can be a costly process involving the computation and normalizing of vectors to multiple light sources and to the viewer, therefore it may become a really slow process if each point in the scene has to be processed. Since quite often the surfaces of the objects in 3D scenes are interpolated with flat polygons, there are a few methods used for interpolating illumination across the polygonal surfaces so that a lesser number of actual illumination computation will be required. Let's study a few of these methods.

5.3.1   Flat Shading

For models defined by collections of polygonal facets or triangles, each facet has a common surface normal. If the light is directional, then the diffuse contribution is constant across the facet. If the eye is also located infinitely far away from the object, and the light is directional, then the specular contribution is constant across the facet. Therefore, the simplest shading method applies only one illumination calculation for each facet (e.g. a triangle). This technique is called *constant* or *flat shading*. It is often used on polygonal shapes. In this method, the illumination is computed for only a single point on each facet. Usually, this is the center point. For a convex facet that has n vertices, the center point is calculated as follows:

$$\mathbf{P}_{centre} = \frac{1}{n} \sum_{i=1}^{n} \mathbf{P}_i \tag{5.4}$$

After that, illumination at this point is to be calculated and used for any other point on this polygon as well. This method is very fast for computation, and it gives good approximation for faceted objects illuminated by directional light sources

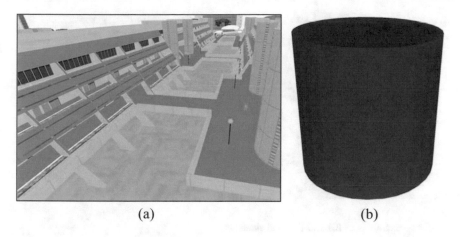

(a)   (b)

Fig. 5.12 Examples of flat-shaded objects

located infinitely far away (Fig. 5.12a). However, for those objects that have curved surfaces, the polygonal nature of the surface interpolation becomes very visible if this method is applied (Fig. 5.12b).

5.3.2   Gouraud Shading

A better appearance can be achieved if illumination values are interpolated from one vertex to another across the surface. For one polygon, it can be done by calculating illuminations at its vertices, and then by interpolating the illumination value between the vertices using the familiar bilinear interpolation (Fig. 5.13a). In this case, the number of illumination calculations is equal to the number of vertices

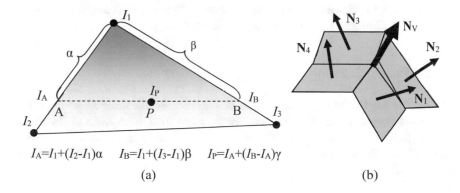

$I_A = I_1 + (I_2 - I_1)\alpha$ $I_B = I_1 + (I_3 - I_1)\beta$ $I_P = I_A + (I_B - I_A)\gamma$

(a)   (b)

Fig. 13 Interpolation of illumination between vertices

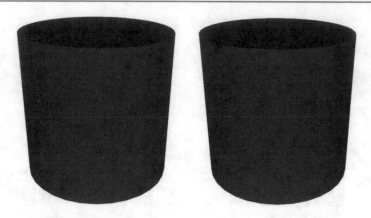

Fig. 5.14 Examples of flat and gouraud shadings

of the polygons used in the scene. This is still a very small value compared to calculating illumination at every point in the scene which is to be equal to the number of pixels of the image. This shading model is called *Gouraud Shading* [3]. When applying this model to a set of adjacent polygons, instead of the actual normals at vertices calculated for different polygons, we have to use average normals calculated from the normals of the polygons which share the vertex (Fig. 5.13b). Alternatively, the normals can be calculated from analytical functions defining the shape, provided these functions are available.

Gouraud shading produces a smoothly shaded polygon mesh so that edges between individual polygons are not visible (Fig. 5.14). This appearance is possible because of the averaged normals and linear interpolation of the color between the vertices. Problems with Gouraud shading may occur when big polygons are used. Imagine we have a large polygon, lit by a light near its center. The light intensity at each vertex will be quite low, because they are far from the light. The polygon will be rendered quite dark, which is wrong, because its center should be brightly lit. However, if we will use a large number of small polygons, with a relatively distant light source, Gouraud shading can look quite acceptable.

5.3.3 Phong Shading

The smaller the size of the polygons, the closer it comes to *Phong Shading* [1] that is a more advanced though computationally more expensive method. In Phong shading (not to be confused with Phong's illumination model), the surface normal is linearly interpolated across polygonal facets, and the illumination model is applied at every point (Fig. 5.15).

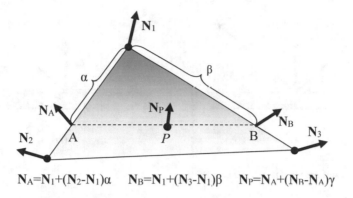

$$N_A = N_1 + (N_2 - N_1)\alpha \qquad N_B = N_1 + (N_3 - N_1)\beta \qquad N_P = N_A + (N_B - N_A)\gamma$$

Fig. 5.15 Phong shading

(a)   (b)

Fig. 5.16 **a** Gouraud and **b** Phong shading

The Phong shading method produces a more accurate calculation of intensity values that creates a more realistic display of surface highlights (Fig. 5.16). However, the Phong method requires significantly more computations than the Gouraud method.

5.4 Shadows and Transparency

When computing colors, we often need to render *shadows* and make transparent objects. To create shadows, the simplest method would be to modify the illumination equation by introducing a shadow coefficient S_i to it:

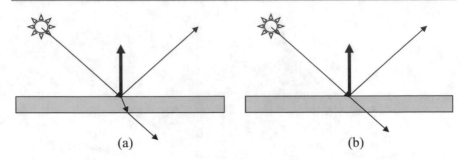

Fig. 5.17 Transparency and refraction

$$I = k_a I_a + \sum_{m}^{lights} S_i I_{L_i} (k_d(\mathbf{N} \cdot \mathbf{L}) + \mathrm{k_s}(\mathbf{V} \cdot \mathbf{R})^n) \qquad (5.5)$$

Then, we have to cast a ray from point P toward each light source L_i. If the ray cannot reach the light source as it is blocked by other objects, then $S_i = 0$, otherwise $S_i = 1$.

To simulate *transparency*, we have to take into account both reflected rays, and the rays coming through the transparent objects (Fig. 5.17a). Generally, when a light beam is incident upon a transparent surface, part of it is reflected and part is transmitted through the material as *refracted light*. The path of refracted light is different from that of incident light. It is a function of the material property (index of refraction). The simplest method for modeling transparent objects is to ignore the path shift due to refraction (Fig. 5.17b). In that case, we can combine the transmitted intensity I_t through a transparent surface from a background object with the reflected intensity I_r from the surface. By introducing an opacity coefficient k_o, the total surface intensity can be calculated as

$$I = I_t + k_o(I_r - I_t) \qquad (5.6)$$

where $k_o = 0$ if the object is translucent; $k_o = 1$ if the object is opaque, and $0 < k_o < 1$ if the object is semi-translucent.

5.5 Textures

Besides colors, real objects have different textures. A common method of adding these surface details to an object is to paint texture patterns onto the surface of the object. The texture pattern may be defined either as an image or an array of color values, or as a procedure that modifies object colors. By *texture mapping*, we refer

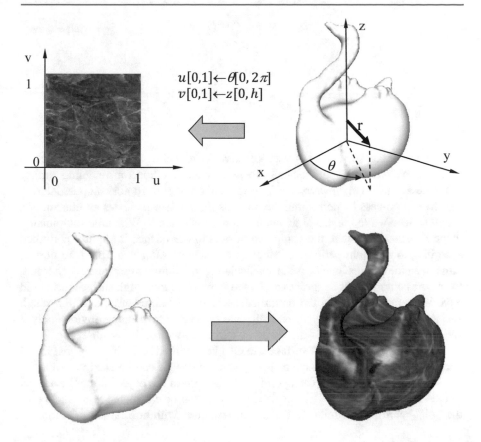

$$u[0,1] \leftarrow \theta[0, 2\pi]$$
$$v[0,1] \leftarrow z[0, h]$$

Fig. 5.18 Cylindrical texture mapping

to a method of incorporating object details into a scene. Texture mapping assumes that the texture image is first either scanned or created. It is then stored in a 2D array of colors, similar to a frame buffer array. Finally, the texture mapping function is to be defined. This function maps coordinates of the surface onto image coordinates. As such, when shading a surface point, we will look up the appropriate pixel from the 2D image to use it to affect the final color. Different texture mapping methods vary in terms of functions used for coordinate mapping.

The example presented in Fig. 5.18 uses so-called *cylindrical texture mapping*. The coordinate mapping function for this example will be as follows:

$$x = r\cos\theta$$
$$x = r\sin\theta$$
$$u = \frac{\theta}{2\pi}$$
$$v = \frac{z}{h}$$
$$\theta \in [0, 2\pi]$$

(5.7)

Mapping images may be not very effective for modeling rough surfaces such as oranges, concrete walls, and asphalt blocks. A better method for modeling surface bumpiness is called *bump mapping*. This method changes the color computed during the shading process by perturbing the normals (Fig. 5.19). It creates an illusion of a rough surface while the actual geometric surface is smooth. When the illumination value is being calculated, the computed normal to the surface is slightly perturbed according to a certain periodic or stochastic function. After that, this fake normal rather than the original one is used for calculating the illumination value. As a result, the colors computed for the point will be the same as for the rough surface that could have the same normals as the perturbed ones. If the bumps are small, the rough surface appearance will be very realistic, and the fake may become obvious only if we examine the boundary of the shape or move really close to its surface.

Finally, the geometry of a surface can also be modified in such a way that it will become rough. Like with image mapping, texture images can be used to define the amount of displacement for each point on the surface of the object along the normal to the surface (Fig. 5.20). This displacement can also be calculated as a function of the point coordinates similar to the way it was done with normal perturbation.

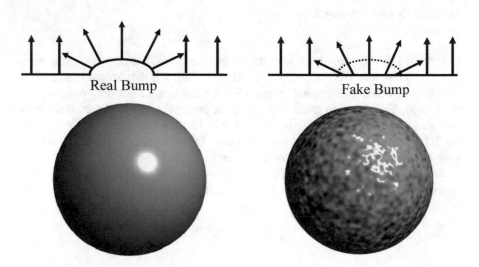

Figure 5.19 Bump mapping

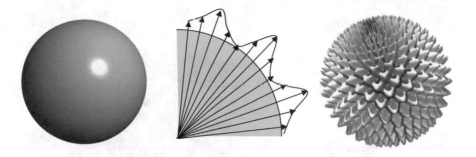

Fig. 5.20 3D texture mapping

5.6 Colors by Functions

Variable colors can be produced procedurally using the very same implicit, explicit and parametric functions that we used for defining geometry of the shapes. This can be done using visualization software that allows for defining colors per vertex in the polygon mesh (e.g. OpenGL and FVRML/FX3D presented in Chap. 7).

For example, to replace a 3D sphere with its image displayed on a plain polygon (called *imposter*), we can use color functions based on the same function that is used for calculating points on the surface of a sphere (Fig. 5.21):

$$\text{Sphere} : R^2 - \left(x^2 + y^2 + z^2\right) = 0$$
$$\text{Color functions} : r = 1 - \min\left((x^2 + y^2 + z^2)/R^2, 1\right), g = 0, b = 0$$

Colors linearly changing from 0 to 1 are displayed on a surface of a 3D cube in Fig. 5.22.

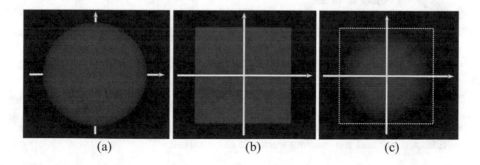

Fig. 21 **a** 3D sphere **b** plain polygon **c** procedural color displayed on a vertical polygon

$$r = \frac{x}{x_{max}}$$
$$g = \frac{y}{y_{max}}$$
$$b = \frac{z}{z_{max}}$$

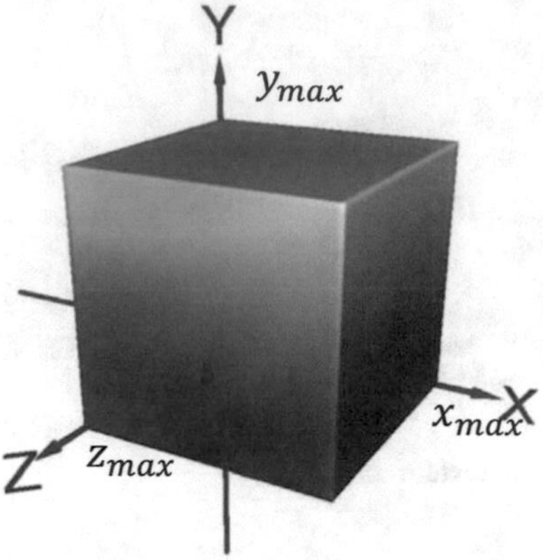

Fig. 22 A 3D color cube

One more example of making a shape with a watermelon look is given in Fig. 5.23. The geometry of the shape is created as a solid sphere with a piece cut of it. The color is defined in the whole 3D space based on computing a color distance function f_c with a noise. The function values are then mapped to the colors linearly interpolated through a few key color patterns. These 3D colors are then sampled on the surface of the geometry placed into this color space. This approach can be implemented using FVRML/FX3D software tools discussed in Chap. 7.

5.7 Summary

- The task of illumination is to determine how the light emitted by light sources reflects from surfaces of shapes and produces what we perceive as color. This task includes lighting and shading. Lighting is a process of computation of the luminous intensity reflected from a specified 3D point. Shading is a process of assigning colors to pixels.
- The common light source types are directional, point, spot, and area lights. The computational complexity of these light sources increases from the directional to the area light sources.
- The empirical illumination model includes ambient, diffuse, and specular reflection parts.
- The brightest point on a surface is where the angle between the normal and the light vector is the smallest.
- The brightest specular reflection is achieved where the angle between the reflected ray and the viewing vector is the smallest.

$$\min(1 - x^2 - y^2 - z^2, -\min(-x + z, \min(y, z)) \geq 0$$

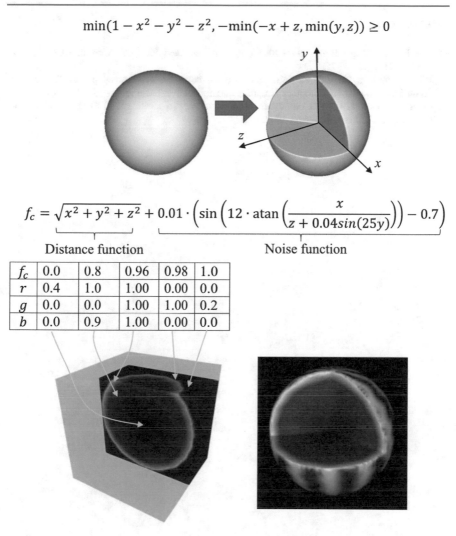

$$f_c = \underbrace{\sqrt{x^2 + y^2 + z^2}}_{\text{Distance function}} + \underbrace{0.01 \cdot \left(\sin\left(12 \cdot \text{atan}\left(\frac{x}{z + 0.04 sin(25y)}\right)\right) - 0.7\right)}_{\text{Noise function}}$$

f_c	0.0	0.8	0.96	0.98	1.0
r	0.4	1.0	1.00	0.00	0.0
g	0.0	0.0	1.00	1.00	0.2
b	0.0	0.9	1.00	0.00	0.0

Fig. 23 Watermelon by functions

- Besides assigning a color to a surface, different texture mappings can be used. The methods of creating textures range from painting textures onto the surface to actual modeling of 3D textures.
- Procedural colors can be defined by using mathematical functions and algorithmic procedures if the visualization software allows for defining colors per vertex in the polygon mesh.

References

1. Bui Tuong Phong, "Illumination for Computer Generated Pictures", Comm. ACM, Vol 18 (6):311–317, June 1975.
2. Blinn, James F., "Models of light reflection for computer synthesized pictures". Proc. 4th Annual Conference on Computer Graphics and Interactive Techniques: 192–198, 1977.
3. Gouraud, Henri (1971). "Continuous shading of curved surfaces". IEEE Transactions on Computers. C-20 (6): 623–629. 1971.

Putting Everything Together

6

6.1 Interactive and Real-Time Rendering

When a computer graphics system is able to react to interaction with the user within 1–3 s, we are working *interactive computer graphics* [1]. When a computer graphics system is able to react within our own time frame, we are dealing with *real-time rendering* [2]. Real-time computer graphics systems are used in computer games, simulation systems, and virtual reality systems. The challenge for such systems is to be able to render large scenes very fast. The observer in a real-time rendering system usually moves through the scene, and this motion can be either predefined or interactively controlled by the user. That means that besides modeling transformations of possible moving objects, we should be able to perform visualization transformation of a relatively small portion of the scene for a moving observer so that the resulting images will be refreshed 60 times per second or even faster. In this chapter, we will learn a few methods of defining the observer position in the World Coordinate System (WCS), as well as learn the common techniques used for fast real-time rendering.

6.2 Positioning the Observer

In Chap. 3, we considered the visualization pipeline where the observer's position was defined in the World Coordinate System. In 2D space, it was a window defined in the WCS, while in 3D space it was a viewing volume. When we anticipate that the observer's position (Fig. 6.1) may change freely, the visualization pipeline has to include one more transformation which is the *viewing transformation*. This transformation maps coordinates from the WCS to the Observer's (or Viewer's) Coordinate System (OCS). Then, in the OCS, the projection transformation will be performed. There are several methods that can be used for defining the OCS within the WCS, and, respectively, for performing the coordinate mapping from WCS to

© The Author(s), under exclusive license to Springer Nature Switzerland AG 2021
A. Sourin, *Making Images with Mathematics*, Undergraduate Topics
in Computer Science, https://doi.org/10.1007/978-3-030-69835-5_6

OCS. They have their advantages and disadvantages and are to be selected
depending on the application problem to be solved.

6.2.1 Direction Cosines

The first method is the method of *direction cosines* [3]. Using this method, the OCS
is defined within the WCS with the three coordinates of its origin, and the direction
cosines of the angles between the coordinate axes of the two coordinate systems
(Fig. 6.2). The coordinate mapping then will be performed with the following
matrices:

$$
\begin{bmatrix} x_{ocs} \\ y_{ocs} \\ z_{ocs} \\ 1 \end{bmatrix} = \begin{bmatrix} c_{11} & c_{12} & c_{13} & 0 \\ c_{21} & c_{22} & c_{23} & 0 \\ c_{31} & c_{32} & c_{33} & 0 \\ 0 & 0 & 0 & 1 \end{bmatrix} \begin{bmatrix} 1 & 0 & 0 & -x_0 \\ 0 & 1 & 0 & -y_0 \\ 0 & 0 & 1 & -z_0 \\ 0 & 0 & 0 & 1 \end{bmatrix} \begin{bmatrix} x_{wcs} \\ y_{wcs} \\ z_{wcs} \\ 1 \end{bmatrix}
\tag{6.1}
$$

where

c_{11}, c_{12}, c_{13} are direction cosines of X_{ocs},

c_{21}, c_{22}, c_{23} are direction cosines of Y_{ocs},

c_{31}, c_{32}, c_{33} are direction cosines of Z_{ocs}.

For example, for the OCS in Fig. 6.3, the matrices will become

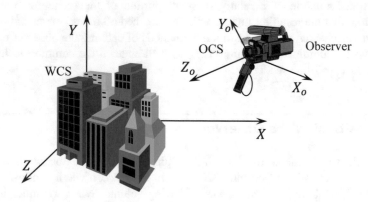

Fig. 6.1 The world coordinate system and the observer's coordinate system

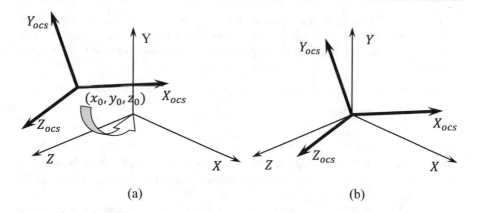

Fig. 6.2 Defining OCS with direction cosines

$$\begin{bmatrix} x_{ocs} \\ y_{ocs} \\ z_{ocs} \\ 1 \end{bmatrix} = \begin{bmatrix} -1 & 0 & 0 & 0 \\ 0 & 0 & 1 & 0 \\ 0 & 1 & 0 & 0 \\ 0 & 0 & 0 & 1 \end{bmatrix} \begin{bmatrix} 1 & 0 & 0 & -5 \\ 0 & 1 & 0 & -3 \\ 0 & 0 & 1 & -1 \\ 0 & 0 & 0 & 1 \end{bmatrix} \begin{bmatrix} x_{wcs} \\ y_{wcs} \\ z_{wcs} \\ 1 \end{bmatrix}$$

$$= \begin{bmatrix} -1 & 0 & 0 & 5 \\ 0 & 0 & 1 & -1 \\ 0 & 1 & 0 & -3 \\ 0 & 0 & 0 & 1 \end{bmatrix} \begin{bmatrix} x_{wcs} \\ y_{wcs} \\ z_{wcs} \\ 1 \end{bmatrix} \tag{6.2}$$

For a moving OCS, this matrix equation turns into the following:

$$\begin{bmatrix} x_{ocs} \\ y_{ocs} \\ z_{ocs} \\ 1 \end{bmatrix} = \begin{bmatrix} c_{11}(t) & c_{12}(t) & c_{13}(t) & 0 \\ c_{21}(t) & c_{22}(t) & c_{23}(t) & 0 \\ c_{31}(t) & c_{32}(t) & c_{33}(t) & 0 \\ 0 & 0 & 0 & 1 \end{bmatrix} \begin{bmatrix} 1 & 0 & 0 & -x_0(t) \\ 0 & 1 & 0 & -y_0(t) \\ 0 & 0 & 1 & -z_0(t) \\ 0 & 0 & 0 & 1 \end{bmatrix} \begin{bmatrix} x \\ y \\ z \\ 1 \end{bmatrix} \tag{6.3}$$

Fig. 6.3 Example of defining OCS with direction cosines

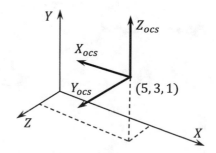

The method of direction cosines is suitable when the cosines rather than the angles themselves are available from the application problem. Also, the direction cosines can be easily calculated as dot products of the respective unit vectors of the coordinate axes.

6.2.2   Fixed Angles

The second method of defining the observer's location is called the method of *fixed angles*. This method originates from the air navigation method where three angles *roll, pitch and yaw* are used to define the orientation of the airplane relative to the fixed coordinate system (Fig. 6.4).

When this method is used, the observer is first rotated about axis Z, then about axes X and Y, and finally translated to the respective 3D point:

$$\mathbf{P}' = \mathbf{Trans}\,\mathbf{Rot}_{yaw}\,\mathbf{Rot}_{pitch}\,\mathbf{Rot}_{roll}\,\mathbf{P} \tag{6.4}$$

Therefore, the inverse transformation will perform the coordinate mapping from the WCS to the OCS:

$$\mathbf{P}_{ocs} = \mathbf{Rot}_{roll}^{-1}\,\mathbf{Rot}_{pitch}^{-1}\,\mathbf{Rot}_{yaw}^{-1}\,\mathbf{Trans}^{-1}\,\mathbf{P}_{wcs} \tag{6.5}$$

The matrices of rotation are as follows:

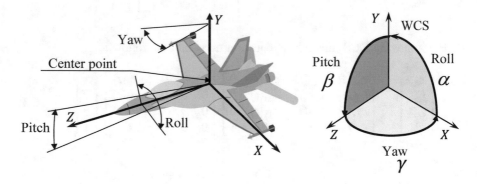

Fig. 6.4 Fixed angles

$$\mathbf{Rot}_{roll}^{-1} = \begin{bmatrix} \cos\alpha & \sin\alpha & 0 & 0 \\ -\sin\alpha & \cos\alpha & 0 & 0 \\ 0 & 0 & 1 & 0 \\ 0 & 0 & 0 & 1 \end{bmatrix}$$

$$\mathbf{Rot}_{pitch}^{-1} = \begin{bmatrix} 1 & 0 & 0 & 0 \\ 0 & \cos\beta & \sin\beta & 0 \\ 0 & -\sin\beta & \cos\beta & 0 \\ 0 & 0 & 0 & 1 \end{bmatrix} \qquad (6.6)$$

$$\mathbf{Rot}_{yaw}^{-1} = \begin{bmatrix} \cos\gamma & 0 & -\sin\gamma & 0 \\ 0 & 1 & 0 & 0 \\ \sin\gamma & 0 & \cos\gamma & 0 \\ 0 & 0 & 0 & 1 \end{bmatrix}$$

Since the angles of rotation and the translation parameters change through time, the matrix equation transforming the world coordinates to the observer's coordinates will be as follows:

$$\mathbf{P}_{ocs} = \mathbf{Rot}_{roll}^{-1}(t)\ \mathbf{Rot}_{pitch}^{-1}(t)\ \mathbf{Rot}_{yaw}^{-1}(t)\ \mathbf{Trans}^{-1}(t)\ \mathbf{P}_{wcs} \qquad (6.7)$$

For the observer location shown in Fig. 9.3, the observer transformation is achieved by *roll* = 180°, *pitch* = −90°, *yaw* = 0° and translation by 5, 3, 1. Therefore, the coordinate transformation from WCS to OCS will be performed as follows:

$$\begin{bmatrix} x_{ocs} \\ y_{ocs} \\ z_{ocs} \\ 1 \end{bmatrix} = \begin{bmatrix} -1 & 0 & 0 & 0 \\ 0 & -1 & 0 & 0 \\ 0 & 0 & 1 & 0 \\ 0 & 0 & 0 & 1 \end{bmatrix} \begin{bmatrix} -1 & 0 & 0 & 0 \\ 0 & 0 & -1 & 0 \\ 0 & 1 & 0 & 0 \\ 0 & 0 & 0 & 1 \end{bmatrix} \begin{bmatrix} 1 & 0 & 0 & 0 \\ 0 & 1 & 0 & 0 \\ 0 & 0 & 1 & 0 \\ 0 & 0 & 0 & 1 \end{bmatrix}$$

$$\begin{bmatrix} 1 & 0 & 0 & -5 \\ 0 & 1 & 0 & -3 \\ 0 & 0 & 1 & -1 \\ 0 & 0 & 0 & 1 \end{bmatrix} \begin{bmatrix} x_{wcs} \\ y_{wcs} \\ z_{wcs} \\ 1 \end{bmatrix} = \begin{bmatrix} -1 & 0 & 0 & 5 \\ 0 & 0 & 1 & -1 \\ 0 & 1 & 0 & -3 \\ 0 & 0 & 0 & 1 \end{bmatrix} \begin{bmatrix} x_{wcs} \\ y_{wcs} \\ z_{wcs} \\ 1 \end{bmatrix} \qquad (6.8)$$

Note that the resulting matrix is identical to the one derived with the direction cosines.

It is important to mention that great care must be taken with these angles when referring to other books and technical papers—the order and meaning of the angles may be different. We used a right-handed coordinate system with the vertical *y*-axis. The sequence for applying the rotations was *roll, pitch* then *yaw*.

The method of fixed angles is very convenient to control the location and motion of the observer defined through the fixed coordinate system. However, if it is the observer who navigates within the scene, the method may cause problems since the observer always has to remember the location of the WCS to measure the angles with reference to this system.

6.2.3 Euler Angles

The third method of locating the observer is called the method of *Euler angles* [3]. In contrast to the method of the fixed angles, the Euler angles, which are also *roll, pitch,* and *yaw,* are measured with reference to the moving observer's coordinate system. Therefore, it is convenient for the observer to navigate within the scene since each time the rotation angles are to be measured with reference to the current location of the OCS axes. Naturally, it is like when we walk. For example, if we need to turn right, we assume right from our current position rather than from some other coordinate systems. In fact, a compound rotation formed by successive *roll, pitch,* and *yaw* rotations about a fixed WCS is equivalent to a compound rotation formed by reversing the sequence of Euler angles. Therefore, we can easily compute a matrix of the compound the Euler angles rotation through the respective fixed angles rotations. If the OCS is located in the WCS using the Euler angles, then coordinates $(x_{wcs}, y_{wcs}, z_{wcs})$ of any point in the WCS transform to coordinates $(x_{ocs}, y_{ocs}, z_{ocs})$ in the OCS by the following matrix transformation:

$$\mathbf{P}_{ocs} = \mathbf{Rot}_{yaw}^{-1} \ \mathbf{Rot}_{pitch}^{-1} \ \mathbf{Rot}_{roll}^{-1} \ \mathbf{Trans}^{-1} \ \mathbf{P}_{wcs} \tag{6.9}$$

where the Euler angles are used in the fixed angles rotation matrices.

For example, for the observer's location shown in Fig. 9.3, the observer's transformation is achieved by *roll* = 180°, *pitch* = 90°, *yaw* = 0° and translation by.

5, 3, 1. Therefore, the coordinate transformation from WCS to OCS will be performed as follows:

$$
\begin{bmatrix} x_{ocs} \\ y_{ocs} \\ z_{ocs} \\ 1 \end{bmatrix} =
\begin{bmatrix} 1 & 0 & 0 & 0 \\ 0 & 1 & 0 & 0 \\ 0 & 0 & 1 & 0 \\ 0 & 0 & 0 & 1 \end{bmatrix}
\begin{bmatrix} 1 & 0 & 0 & 0 \\ 0 & 0 & 1 & 0 \\ 0 & -1 & 0 & 0 \\ 0 & 0 & 0 & 1 \end{bmatrix}
\begin{bmatrix} -1 & 0 & 0 & 0 \\ 0 & -1 & 0 & 0 \\ 0 & 0 & 1 & 0 \\ 0 & 0 & 0 & 1 \end{bmatrix}
$$
$$
\begin{bmatrix} 1 & 0 & 0 & -5 \\ 0 & 1 & 0 & -3 \\ 0 & 0 & 1 & -1 \\ 0 & 0 & 0 & 1 \end{bmatrix}
\begin{bmatrix} x_{wcs} \\ y_{wcs} \\ z_{wcs} \\ 1 \end{bmatrix} =
\begin{bmatrix} -1 & 0 & 0 & 5 \\ 0 & 0 & 1 & -1 \\ 0 & 1 & 0 & -3 \\ 0 & 0 & 0 & 1 \end{bmatrix}
\begin{bmatrix} x_{wcs} \\ y_{wcs} \\ z_{wcs} \\ 1 \end{bmatrix} \tag{6.10}
$$

Note that we have got exactly the same matrix again. Whatever method is used for positioning the observer in the WCS, the resulting transformation from the WCS to the OCS is to be the same.

6.3 Fast Rendering of Large Scenes

Now that we learned how to map coordinates from the WCS to the OCS, let's consider how to organize shape models to be able to perform this coordinate transformation as well as the whole visualization transformation as fast as possible.

6.3.1 Hierarchical Representation

Large scenes involve complex shapes which may consist of similar parts like buildings which may consist of similar or identical blocks. Therefore, geometric models often have a hierarchical structure induced by a bottom-up construction process: components are used as building blocks to create higher level entities which in turn serve as building blocks for yet higher level entities, and so on. These components are called *modules*. The lowest level of hierarchy is represented by *primitive instances* or *primitive geometric objects* or just *primitives*. This hierarchical relationship can be defined either by a *tree* or by a *Directed Acyclic Graph (DAG)*.

The tree is a rather uncommon case when each object is included only once in a higher level object (Fig. 6.5), while the DAG is a more common case of objects included multiple times (Fig. 6.6).

The basic primitives must be defined in independent coordinate systems which are then transformed to the world coordinates. These systems are called *master coordinate systems*. Any transformations applied to the master coordinate definition of a primitive are called *modeling transformations*. The creation of an *instance* in the world coordinates from a master coordinate primitive definition is called an *instance transformation*.

Using DAGs allows for the construction of complex objects in a modular fashion by repetitive invocation of objects that vary in geometric and appearance attributes. As a result, the storage economy increases since it is sufficient to store only references to objects that are used repeatedly. It also allows for easy update

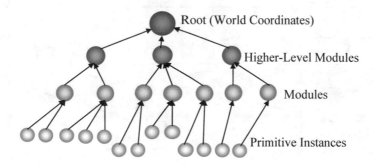

Fig. 6.5 Hierarchy representation with a tree

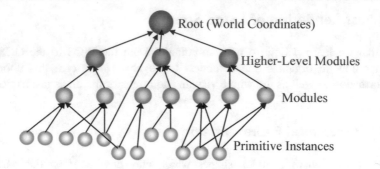

Fig. 6.6 Hierarchy representation with the directed acyclic graph

propagation, because a change in the definition of one building-block automatically propagates to all higher level objects that use that object. Finally, it speeds up the visualization.

In terms of modeling transformations, the DAG hierarchy can be symbolized as it is shown in Fig. 6.7.

6.3.2 Current Transformation Matrix

We can see that with each node we can associate a composite transformation matrix which will perform an instance transformation from the respective level of hierarchy to the root level of the World Coordinate System (Level$_4$ in our example). A

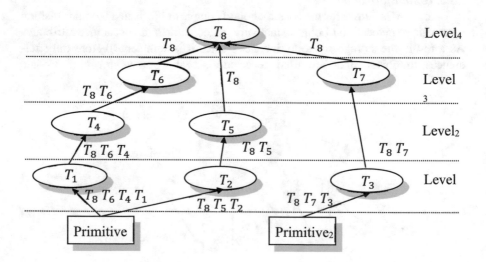

Fig. 6.7 Symbolization of hierarchy with modeling transformations

Current Transformation Matrix (CTM) for any given node in DAG is the composed matrix transformation accumulated to perform the instance transformation for this node. As we traverse down the DAG, we multiply the CTM with the lower level transformation matrix for column position vector representation or premultiply the CTM for row representation. Using CTM significantly accelerates the rendering speed since we no longer need to compute the transformation matrices each time an instance transformation is to be performed.

6.3.3 Logical and Spatial Organizations

There can be two different ways of implementing the hierarchical organization of large data scenes in DAG. The first one is called *Logical Organization*, and usually it is used while the scene is being designed and built. In this organization, the nodes are logical modules which combine objects of the same type. It allows for easy identification of individual parts of the scene for editing purposes.

For example, in Fig. 6.8, all the trees are combined under the "Trees" node, and they are created from one primitive "Tree" by applying different affine transformations T_i. Inclusion of any new instance of the existing primitives is carried out by adding to the respective node of one more transformation node making a new instance (e.g. T_{11} for adding a new house). Deletion of an instance is to be done similarly by deleting the respective node. Should we need to change the appearance of the whole group of nodes, we only need to replace the respective primitive(s).

Though easy to operate with, this organization is not really efficient when the scene is to be visualized. This is because when performing visualization, only a little portion of the large scene is to be visible and respectively sent to the

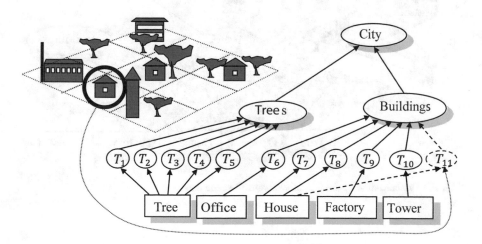

Fig. 6.8 Logical organization of the scene's database

visualization pipeline, while the logical organization will require the graphics system to traverse the whole DAG to check which of the shapes will be visible (Fig. 6.9).

This deficiency can be overcome if the scene database is organized differently. If instead of the logical nodes, the instances of the primitives will be associated with spatial tiles (e.g. a regular grid), then we have *Spatial Organization* of the scene's database (Fig. 6.10). Like for the logically organized database, inclusion or deletion of an instance will require addition or removal of a transformation node. This node will be associated with the respective tile where the instance of the primitive shape is supposed to appear within.

For the spatially organized scene's database, the visualization speed can be significantly improved, since only those nodes which are associated with the tiles visible to the observer (Fig. 6.11) will be rendered.

The task of determining the potentially visible tiles is incomparably faster than task of rendering of all the nodes, therefore, the whole rendering will be performed much faster. Of course, it is rather difficult to organize the scene spatially. Therefore, usually when the scene is being created, logical organization is used, and only when the design is complete, the scene database is converted to the spatial organization. Normally, special software tools are used for it, for which only the size of a regular grid is to be defined to get the resulting database organized spatially. It is done in a manner of slicing a cake into pieces, therefore some of the shapes, occupying several tiles, may be split up into several adjoining shapes.

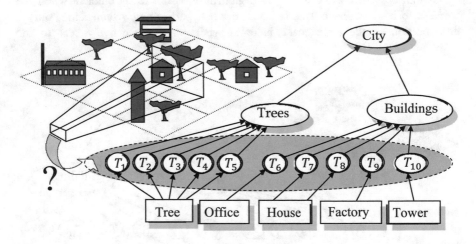

Fig. 6.9 Visualization of the logically organized scene database

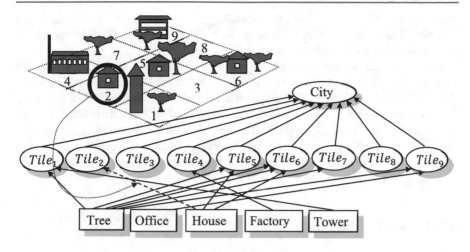

Fig. 6.10 Spatial organization of the scene's database

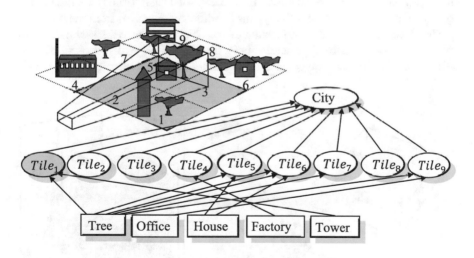

Fig. 6.11 Visualization of the spatially organized scene's database

6.3.4 Bounding Boxes

Spatial organization assumes the fact that before attempting to render all the shapes, we first determine which of the areas are potentially visible. This technique can be further extended to individual shapes by introducing so-called *bounding boxes*.

The bounding boxes are used in the scene to define the location and overall size of the shapes they are associated with. The bounding boxes are not displayed, however, they are parts of the database and therefore participate in rendering. When the bounding boxes are in use, the rendering software first checks which of the bounding boxes are potentially visible. After that it performs rendering of the shapes associated with the visible bounding boxes (Fig. 6.12).

Since the shapes defined within the bounding boxes are expected to have a lot more polygons than the boxes themselves, the overall rendering speed will be faster since only the potentially visible shapes will be rendered. The method of bounding boxes can also be efficiently used when the problem of collision detection is used in virtual reality and computer games software. Besides boxes, other simple shapes like spheres and cylinders can be used as bounding shapes. As soon as the enclosed shape has a much higher complexity than the bounding shape, it will allow for increasing the rendering performance.

6.3.5   Level of Detail

The next method of accelerating rendering of a large scene is using the *level-of-detail*. When using this method, several representations of the same shape are stored in the scene's database. Each Level-of-Detail (LOD) representation consists of a different number of polygons for different distances from the observer.

Therefore, LOD for a distant location of the object will have a lesser number of polygons, while for closer distances more detailed LODs will be used. This is because highly detailed polygonal models cannot really be seen in all the details

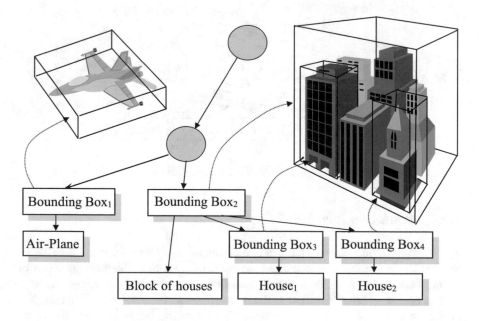

Fig. 6.12 Bounding boxes

when the distance from the observer is large. It is illustrated in Fig. 6.13, where a cone has three LODs defined for three different distances of the cone from the observer. After a certain distance, the object may not be displayed at all.

When polygonal models are stored in the database, several LODs will be associated with the respective node rather than just one model (Fig. 6.14). Only one LOD will be used at each time depending on the distance from the observer to the object. Though computation of a distance takes a certain time, it compensates with a much lesser time for visualizing simplified models in place of highly detailed models when the distance from the observer is such that these details cannot be seen.

When the graphics system switches between LODs, the image will unnaturally jump between two representations since they visually have a different number of polygons. To avoid it, blending between two polygonal shapes can be used, like the morphing considered in Chap. 3 (See Fig. 6.15).

Alternatively, a continuous level-of-detail based on parametric or implicit representation of the shape can be implemented in place of a fixed number of LODs.

Let's consider an example of implementing the continuous level-of-detail parametrically. Let the shape be defined with the following formulas:

$$x = 1.6 \left(\cos(0.5\varphi)\right)^3$$
$$x = 1.6 \left(\cos\theta \sin(0.5\varphi)\right)^3$$
$$z = 1.6 \left(\sin\theta \sin(0.5\varphi)\right)$$
$$\varphi \in [0, 2\pi], \quad \theta \in [0, 2\pi]$$

$$(6.11)$$

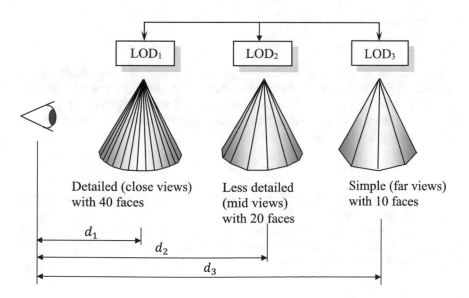

Detailed (close views) with 40 faces

Less detailed (mid views) with 20 faces

Simple (far views) with 10 faces

Fig. 6.13 Level-of-detail $(d_1 < d_2 < d_3)$

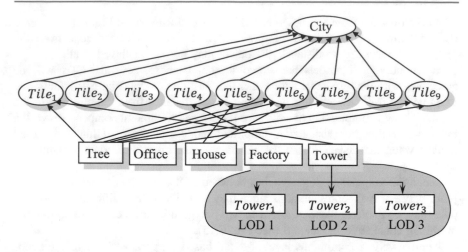

Fig. 6.14 Using the level-of-detail in the spatially organized scene database

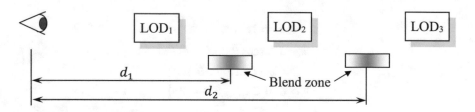

Fig. 6.15 Blending the level-of-details

The number of polygons in its polygonal representation is defined by the increment δ for the two parameters. The smaller δ is, the more polygons will be calculated. The larger δ is, the greater the size of the polygons will be, and, respectively, the lesser polygons will be created. Therefore, a simple linear function of the distance from the observer will do the trick:

$$\delta = k \cdot distance \tag{6.12}$$

In Fig. 6.16, four different polygonal representations generated for four different values of δ are displayed.

Implicit functions also can be used for generating a continuous level-of-detail, however, we will not discuss it here since the details of the algorithm of rendering implicitly-defined shapes are beyond the scope of this book.

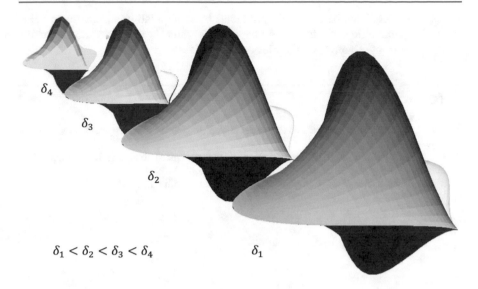

$$\delta_1 < \delta_2 < \delta_3 < \delta_4 \qquad \delta_1$$

Fig. 6.16 Continuous level-of-detail of a shape defined parametrically

6.4 Summary

- When the observer moves in a large 3D scene, we have to perform a viewing transformation which will map the coordinates from the World Coordinate System into the Observer's Coordinate System. Projection transformation will then be performed using the coordinates defined in the OCS.
- *Direction cosines* may not be a convenient way of locating the observer position since we usually operate with angles rather than their cosines. *Fixed angles* are suitable when the observer's position is being defined in the WCS, but for the observer these angles may be difficult to navigate with. For the observer, it is more convenient to navigate with the *Euler angles*. Whatever method is used, the resulting transformation matrix (from the WCS to the OCS) will be the same.
- It is important to apply the *Roll, Pitch,* and *Yaw* angles in a correct order for the *Fixed Angles* and the *Euler Angles*.
- The following methods accelerate visualization of large scenes:

  - Hierarchical representation of a scene and the use of modules.
  - Storing a current transformation matrix for different nodes in the model.
  - Using spatial organization of a scene.
  - Bounding boxes (volumes).
  - Level-of-detail.

- Level-of-detail can be efficiently produced by manipulating an increment value in the parametric definition of a shape. This value is to be a function of the distance to the shape from the observer.

References

1. Angel, E., Shreiner, D., Interactive Computer Graphics. A Top-Down Approach with WebGL, Pearson, 2015.
2. Watt, A., Policarpo, F., 3D Games. Animation and Advanced Real-time Rendering, vol. 2, Addison-Wesley, 2003.
3. Vince, J., Mathematics for Computer Graphics, Springer, 5th ed., 2017

Let's Draw

7

7.1 Programing Computer Graphics and Visualization

When we work on computer graphics problems, it is usually about achieving perfection of rendering points, lines, and polygons while data visualization problems are about how to represent graphically various data which may have no obvious graphical appearance at all.

The software tools available for working on these problems may be given to us as software libraries which are extensions of common programming languages, as well as specially developed stand-alone graphics libraries and interactive systems.

All the graphics and visualization software tools can be also classified by the main programing paradigm which they follow: Imperative or Declarative Style. The imperative style of programming tells the computer *how to do* things, while when the declarative style of programming tells the computer *what to do*.

In this chapter, we will learn how to use a few software tools which are commonly used for solving various graphics and visualization problems: OpenGL which is based on using the imperative polygon-based visualization style, POV-Ray which is declarative style ray tracing software and VRML/X3D which is declarative style polygon-based software. Two more software tools from the author of this book—FVRML/FX3D and Shape Explorer—will be also presented.

7.2 Drawing with OpenGL

7.2.1 Introduction to OpenGL

OpenGL is a software interface to graphics hardware [1]. OpenGL is designed as a hardware-independent interface to be implemented on many different hardware platforms. To achieve it, OpenGL does not have any commands for performing

© The Author(s), under exclusive license to Springer Nature Switzerland AG 2021   177
A. Sourin, *Making Images with Mathematics*, Undergraduate Topics
in Computer Science, https://doi.org/10.1007/978-3-030-69835-5_7

windowing tasks or obtaining user input. Instead, the windowing system controls the particular hardware that has to be used.

In this chapter, we will use only basic OpenGL interface as it was defined in OpenGL version 1.1 since it is enough to illustrate all the concepts described in the previous chapters. It has about 120 distinct commands which can be used for defining 2D/3D points, lines, and polygons. It also supports affine, orthographic, and perspective projection transformations. Other features include RGBA and color-index display modes, multiple light sources, blending, antialiasing, fog, bitmap operations, texture mapping, and multiple frame-buffers.

Each OpenGL command has the following syntax:

$$glCommandSfx(\ldots);$$

where Sfx denotes one of the following suffixes defining the parameters' data type: b – signed char, s – short, i – long, f – float, d – double, ub – unsigned char, us – unsigned short, ui – unsigned long.

OpenGL is a so-called state machine. It can be put into various states that then remain in effect until they are changed again. Each state or mode has a default value. Each state can be queried at any point.

A simple OpenGL code is given in Fig. 7.1. It instructs to draw a white square onto a black background.

In this example, glClearColor sets the cleaning color to black followed by glClear which clears the entire window to the clearing color. Then, glClearDepth and glClear clear the depth buffer. Function glColor3f (red, green, blue) sets the current color, where red, green, blue are of type *float* in the domain [0.0, 1.0]. For example, yellow will be defined with glColor3f(1.0, 1.0, 0.0). Next, lines and polygons have to be defined between calling two functions: glBegin and glEnd The parameter of glBegin (GL_PrimitiveType) defines how exactly to interpret the vertices defined there (Fig. 7.2).

```
glClearColor(0.0, 0.0, 0.0, 0.0);
glClear(GL_COLOR_BUFFER_BIT);
glColor3f(1.0, 1.0, 1.0);
glOrtho(-1.0, 1.0, -1.0, 1.0, -1.0, 1.0);
glBegin(GL_POLYGON);
    glVertex2f(-0.5, -0.5);
    glVertex2f(-0.5, 0.5);
    glVertex2f(0.5, 0.5);
    glVertex2f(0.5, -0.5);
glEnd();
glFlush();
```

Fig. 7.1 A simple OpenGL program

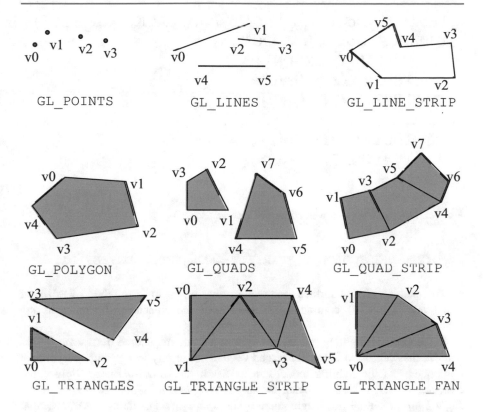

Fig. 7.2 Points, lines, and polygons. Note the order of how the vertices are listed

For polygons, a visible side is to be defined which can be the front, back, or both sides. This can be done with glPolygonMode(face, mode).

Color can be specified either in *RGB* terms with glColor3f(r, g, b), or in *RGBA* terms with glColour4f(r, g, b,] a), or in color-index mode with glIndex(c). The fourth component a in *RGBA* mode is used for defining the opacity of the objects.

Two shading modes can be used: flat shading and smooth shading. If smooth shading of polygons is required, normal vectors (normals) at each vertex have to be provided. The OpenGL code for such a polygon will change to the one listed in Fig. 7.3.

When drawing a scene composed of three-dimensional objects, some of them might obscure others or parts of them. To draw a realistic scene, these obscured parts of the objects must be not displayed. The elimination of such obscured parts of solid objects is called hidden-surface removal. For this purpose, the depth buffer is used in OpenGL. Initially, this buffer is to be set to the initial state using the glClear() command with GL_DEPTH_BUFFER_BIT. This can be done concurrently with clearing the color buffer:

```
glPolygonMode(GL_FRONT_AND_BACK, GL_FILL);
glShadeModel(GL_SMOOTH);
glBegin(GL_POLYGON);
    glNormal3fv(nv0);
    glVertex3fv(vv0);

. . . . . . . . . . . .

    glNormal3fv(nv9);
    glVertex3fv(vv9);
glEnd();
```

Fig. 7.3 Defining a polygon with normals at each vertex

```
glClear(GL_COLOR_BUFFER_BIT | GL_DEPTH_BUFFER_BIT);
```

To enable hidden-surface removal, you have to use the `glEnable` (`GL_DEPTH_TEST`) command, and then draw the objects in the scene in any order.

In the OpenGL lighting model, the light in a scene may come from several light sources that can be individually turned on and off at any time. Four independent components of the lighting model can be used: *Emissive, Ambient, Diffuse*, and *Specular Lights*. For each object, the respective *Material Color* is to be defined. *Light Sources* can be point light sources and spotlights. For them, *color, position, attenuation*, and *direction* are to be specified.

If you are writing a program that you want to work properly both with or without a network, include a call to `glFlush` at the end of each frame or scene. It forces previously issued OpenGL commands to begin execution, thus guaranteeing their completion in finite time.

The whole visualization pipeline of OpenGL is illustrated in Fig. 7.4. Besides defining vertices of points, lines and polygons, it includes Modeling, Projection and Viewport transformations.

Modeling transformation, as we discussed in Chap. 9, is represented in OpenGL by three affine transformations: translation `glTranslate(x,y,z)`, rotation about vector $[x_0\ y_0\ z_0]$ `glRotate(angle, × 0,y0,z0)` and scaling relative to the origin `glScale(sx,sy,sz)`.

Projection can be performed with orthographic projection transformations `glOrtho`, `glOrtho2D`, and perspective projection transformation `ProjectionglFrustum`.

Finally, the viewport transformation is performed with `glViewport` which defines a rectangular viewing area within a window on the screen.

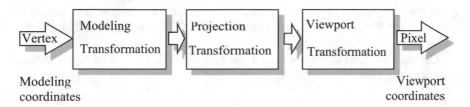

Fig. 7.4 Visualization pipeline of OpenGL

7.2.2 Interaction with GLUT

OpenGL does not do any window management, event handling, or other OS specific functions. It is to be done by the user and other libraries working together with OpenGL. One of such libraries is *GLUT*, originally developed by *Mark Kilgart* and ported to MS Windows by *Nate Robins* [2]. GLUT—the OpenGL Utility Toolkit—is a window system independent toolkit for writing interactive OpenGL programs. There is also a newer version of GLUT called FreeGLUT [3] which supports the recent versions of OpenGL. We, however, will still use GLUT since we only need it for a few very basic operations.

GLUT has about 90 different functions such as Initialization, Handling Input Events, Window Management, Menu Management, Callback Registration, Color Index, Colormap Management, State Retrieval, Font Rendering, and Advanced Geometric Object Rendering.

The most frequently used functions are the following:

`glutInit()` initializes the GLUT library and negotiates a session with the window system.

`glutInitDisplayMode()` sets the initial display mode such as *RGB, RGBA, INDEX, SINGLE,* or *DOUBLE*.

`glutInitWindowSize()`, `glutInitWindowPosition()` set the initial window size and position, respectively.

`glutCreateWindow()` creates a top-level window with a defined name.

`glutMainLoop()` enters the GLUT event processing loop.

`glutDisplayFunc()` sets the display callback function for the current window that redraws the objects in your scene.

`glutIdleFunc()` sets a function that is to be executed if no other events are pending (for example, when the event loop would be idle).

`glutReshapeFunc()` indicates what action should be taken when the window is resized, moved, or exposed.

`glutKeyboardFunc()`, `glutSpecialFunc` and `auxMouseFunc()` allow you to link a keyboard key or a mouse button with a routine that is invoked when the key or mouse button is pressed or released.

GLUT also supports a double-buffering mode for animation where two complete color buffers—the viewable and drawable buffers—are used. Special glutSwapBuffers() routine waits until the current screen refresh period is over so that the previous buffer is completely displayed before swapping the buffers.

The detailed functionality of GLUT [4] and OpenGL [5] can be found in their programming manuals.

7.2.3 Drawing Parametric Shapes with OpenGL

We will use the version of OpenGL shipped with *MS Visual Studio 2017*. However, the OpenGL code will remain the same when you port it to other compilers and hardware platforms.

First, you will need to install GLUT on your computer, since it is not a part of *MS Visual Studio*. Follow these instructions:

1. Download and unzip *GLUT* binary files from the *GLUT* webpage [3].
2. Copy *glut.h* to *C:\Program Files\Microsoft Visual Studio\VC98\Include\GL*
3. Copy *glut.h* to *C:\Program Files (×86)\Microsoft Visual Studio\2017\Enterprise\VC\Tools\MSVC\14.16.27023\include\GL*
4. Copy *glut32.lib* to *C:\Program Files (×86)\Microsoft Visual Studio\2017\Enterprise\VC\Tools\MSVC\14.16.27023\Lib*
5. Copy *glut32.dll* to *C:\Windows\System32*

Now, we will learn how to display with OpenGL curves, surfaces, and solid objects. Copy to your directory the files *OpengGL/Curve* from the book's software repository [6]. You can use the solution *curve.sln* or, if you want to start a new project, open a new project "Win32 Console Application" from the directory of *OpengGL/Curve* and then build and run it. You will see on the screen a graphics application window with a curve displayed there (Fig. 7.5).

Using keys ←, →, ↑, and ↓ you may rotate the scene horizontally and vertically. By pressing "Esc", you will terminate the program.

The code of the functions displaying the coordinate axes and the curve is shown in Fig. 7.19. These functions are only a "tip of the iceberg", and we will look at other supporting functions later in this section.

To display a surface shown in Fig. 7.7, which is obtained from the curve by its rotational sweeping, we change coordinate computation in Function that is now has to be done by functions of two parameters u and v (Fig. 7.8). Then, we have to sample it in Surface_uv by computing triangle strips (Fig. 7.9). This is done in the book software repository project *OpengGL/Surface*. The order of computation of vertices is important here, and after defining the first two vertices of the first triangle in the strip, other triangles are formed by computing coordinates of one vertex at the time. Normal vectors built at the vertices of triangles are computed in function Normal_uv (Fig. 7.10) as cross products of two vectors built at each vertex.

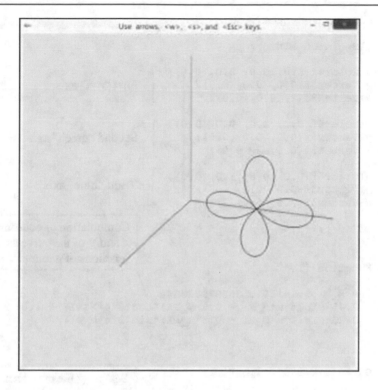

Fig. 7.5 Displaying a curve with the OpenGL/GLUT application. Refer to the project *OpengGL/Curve/*

Generally, to calculate the normal at a given vertex, we have to differentiate the three parametric functions f_1, f_2, and f_3 by the variables u and v which will give us two vectors $\mathbf{n_1}$ and $\mathbf{n_2}$ tangent to the surface in the directions u and v (Fig. 7.11). Then, we have to compute the cross product $\mathbf{n_1} \times \mathbf{n_2}$ which yields us a normal vector orthogonal to both $\mathbf{n_1}$ and $\mathbf{n_2}$ and hence to the surface.

The functions can be differentiated either by computing partial derivatives, or, as it is done in function `Normal_uv`, by using small increments δ_u and δ_v of the parameters u and v:

$$\frac{\partial f_i}{\partial u} = \frac{f_i(u_i, v_i) - f(u_i + \delta_u)}{\delta_u}, \frac{\partial f_i}{\partial v} = \frac{f_i(u_i, v_i) - f(v_i + \delta_v)}{\delta_v}$$

Given two vectors $\mathbf{n}_1 = [x_{n1}\ y_{n1}\ z_{n1}]$ and $\mathbf{n}_2 = [x_{n2}\ y_{n2}\ z_{n2}]$, the cross product is then calculated:

$$\mathbf{N} = [y_{n1}\,z_{n2} - y_{n2}\,z_{n1} \quad x_{n2}\,z_{n1} - x_{n1}z_{n2} \quad x_{n1}\,y_{n2} - x_{n2}\,y_{n1}]$$

```
void CoordinateSystem(void)
{
    glBegin(GL_LINES);

    glColor4f(1.0, 0.0, 0.0, 0.0);  ⎤
    glVertex3f(0.0, 0.0, 0.0);       ⎬  First "red" axis
    glVertex3f(2, 0.0, 0.0);         ⎦

    glColor4f(0.0, 1.0, 0.0, 0.0);  ⎤
    glVertex3f(0.0, 0.0, 0.0);       ⎬  Second "green" axis
    glVertex3f(0.0, 2, 0.0);         ⎦

    glColor4f(0.0, 0.0, 1.0, 0.0);  ⎤
    glVertex3f(0.0, 0.0, 0.0);       ⎬  Third "blue" axis
    glVertex3f(0.0, 0.0, 2);         ⎦

    glEnd();
}
```

> Computation of coordinates x and y of the curve as functions of parameter u

```
void Function(float u)
{
# define M_PI 3.14159265358979323846
    x = 0.75*sin(4*u*M_PI - 0.5*M_PI)*cos(2*u*M_PI) + 1.0;
    y = 0.75*sin(4*u*M_PI - 0.5*M_PI)*sin(2*u*M_PI);
    z = 0;
}

void Curve(float u1, float u2, int nu)
{
    int i;
    float u;
    glBegin(GL_LINE_STRIP);
    glColor4f(0.0, 0.0, 0.0, 0.0);
```

> Sampling the curve nu times and connecting the computed points with straight line segment

```
    for (i = 0; i <= nu; i++) {
        u = u1 + (u2 - u1)*(float)i / (float)nu;
        Function(u);
        glVertex3f(x, y, z);
    }
    glEnd();}
```

Fig. 7.6 Functions displaying the coordinate axes and the curve

For the vector to be normalized, we must bypass special degenerate cases where the normal has zero length to avoid artifacts on the surface (black or white spots). For example, for a parametrically defined sphere:

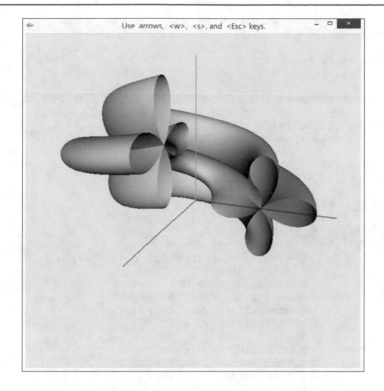

Fig. 7.7 Displaying a surface. Refer to the project *OpengGL/Surface/*

> Computation of coordinates *x y* and *z* of the surface as functions of parameters *u* and *v*

```
void Function(float u, float v)
{
# define M_PI 3.14159265358979323846
    x = (0.75*sin(4*u*M_PI - 0.5*M_PI)*cos(2*u*M_PI) + 1.0)
            *sin(v*3*M_PI/2 + M_PI/2);
    y =   0.75*sin(4*u*M_PI - 0.5*M_PI)*sin(2*u*M_PI)+1.0*v;
    z = (0.75*sin(4*u*M_PI - 0.5*M_PI)*cos(2*u*M_PI) + 1.0)
            *cos(v*3*M_PI/2 + M_PI/2); }
```

Fig. 7.8 Computing coordinates of the surface

```
void Surface_uv(float u1, float u2, float v1, float v2, int nu,
int nv)
{
    float u, v, du,dv;
    int i, j;

    if (RendMode==1){glPolygonMode(GL_FRONT_AND_BACK,GL_LINE);
                     glColor4f(1.0, 1.0, 1.0, 0.0);}
    else {glEnable(GL_LIGHTING);
          glPolygonMode(GL_FRONT_AND_BACK,GL_FILL);}

    du = (u2 - u1) / (float)nu;
    dv = (v2 - v1) / (float)nv;

  for (j = 0; j < nv; j++) {
    v = v1+(v2-v1)*(float)j / (float)nv;
    glBegin(GL_TRIANGLE_STRIP);
    for (i = 0; i <=nu; i++) {
                u = u1+(u2-u1)*(float)i / (float)nu;
                if (i == 0) {
                Function(u, v+dv,w);
                Normal_uv(u, v+dv, du,dv,v2);
                glNormal3f(xn, yn, zn);
                glVertex3f(x, y, z);

                Function(u, v, w);
                Normal_uv(u, v, du, dv,v2);
                glNormal3f(xn, yn, zn);
                glVertex3f(x, y, z);
                                        }
                else {
           Function(u , v + dv, w);
           Normal_uv(u, v + dv, du, dv,v2);
           glNormal3f(xn, yn, zn);
           glVertex3f(x, y, z);

           Function(u, v ,w);
           Normal_uv(u, v, du, dv,v2);
           glNormal3f(xn, yn, zn);
           glVertex3f(x, y, z);
                            }

            }
    glEnd();
  }

}
```

Fig. 7.9 Function displaying the surface

```
void Normal_uv(float u, float v, float du, float dv, float v2)
{
       float nd;
       float x0, y0, z0, x1, y1, z1, x2, y2, z2, du1, dv1;
       x0 = x; y0 = y; z0 = z;
       du1 = du / 10.0;
       dv1 = dv / 10.0;
       if (v != v2) {
            Function(u + du1, v + dv1);
            x1 = x - x0, y1 = y - y0, z1 = z - z0;
            Function(u, v + dv1);
            x2 = x - x0, y2 = y - y0, z2 = z - z0;
                      }else
            {Function(u-du1, v- dv1);
            x1 = x - x0, y1 = y - y0, z1 = z - z0;
            Function(u + du1, v - dv1);
            x2 = x - x0, y2 = y - y0, z2 = z - z0;}
       xn = y1*z2 - y2*z1;
       yn = x2*z1 - x1*z2;
       zn = x1*y2 - x2*y1;

       nd = sqrt(xn*xn + yn*yn + zn*zn);
       x = x0; y = y0; z = z0;
       xn = xn / nd; yn = yn / nd; zn = zn / nd;

}
```

Fig. 7.10 Function computing the normal vectors at the vertices

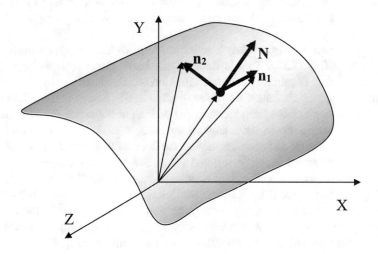

Fig. 7.11 Calculating the normal vector

$$x = r \cos \phi \cos \theta$$
$$y = r \cos \phi \sin \theta$$
$$z = r \sin \phi$$
$$\phi \in \left[-\frac{\pi}{2}, \frac{\pi}{2}\right], \quad \theta \in [-\pi, \pi]$$

the partial derivatives by ϕ and θ are

$$
\begin{aligned}
x_1 &= -r \sin \phi \cos \theta & x_2 &= -r \cos \phi \sin \theta \\
y_1 &= -r \sin \phi \sin \theta & y_2 &= r \cos \phi \cos \theta \\
z_1 &= r \cos \phi & z_2 &= 0 \\
\phi \in \left[-\frac{\pi}{2}, \frac{\pi}{2}\right], & \theta \in [-\pi, \pi] & \phi \in \left[-\frac{\pi}{2}, \frac{\pi}{2}\right], & \theta \in [-\pi, \pi]
\end{aligned}
$$

For the parameter values $\phi = 0$ and $\theta = 0$, the coordinates of the tangent vectors n_1 and n_2 are $x_1 = 0$, $y_1 = 0$, $z_1 = 1$ and $x_2 = 0$, $y_2 = 0$, $z_2 = 0$ which yields a degenerate normal $N = [0\ 0\ 0]$ while it is to be $[1\ 0\ 0]$ at this point:

$$
\begin{bmatrix} i & j & k \\ 0 & 0 & 1 \\ 0 & 0 & 0 \end{bmatrix} = i \begin{bmatrix} 0 & 1 \\ 0 & 0 \end{bmatrix} - j \begin{bmatrix} 0 & 1 \\ 0 & 0 \end{bmatrix} + k \begin{bmatrix} 0 & 0 \\ 0 & 0 \end{bmatrix} = i0 - j0 + k0
$$

If this degenerate normal vector is used, the resulting image will have a black spot at the respective point of the shape (Fig. 7.12) since the illumination at this point has been calculated wrongly. The black color computed at one vertex linearly interpolates to red color across surfaces of the triangles connected to this vertex. These cases of possible degenerate normal are avoided in the code of `Normal_uv`.

Finally, we will display a solid object obtained from the surface by filling in the space inside it (Fig. 7.13). This is done in the book software repository project *OpengGL/Solid*. It requires to redefine function `Function` so that it computes coordinates x, y, z as functions of three parameters u, v, and w (Fig. 7.14). The solid object is then created from the surfaces which are bounding it. With the three parameters, it can be up to six bounding surfaces, e.g. six faces of a cube. We, therefore, can reuse the function displaying surfaces so that it will display surfaces defined by parameters (u, v), (v, w), and (w, u) for the two respective extreme values of parameters w, u, and v. This is done in function `Solid` (Fig. 7.15) which is calling `Surface_uv` (Fig. 7.16) as well as `Surface_vw` and `Surface_wu` obtained from `Surface_uv` by coordinate permutations. These three functions use function `Normal_uv`, as well as `Normal_uv` and `Normal_uv` which are obtained from `Normal_uv` by coordinate permutations (Fig. 7.17).

To make all these functions work, there is a significant overhead of additional functions responsible for setting OpenGL environment and interaction using GLUT. Thus, Function `Init` (Fig. 7.18) is called only once before any other functions are called. It defines parameters of the light source `LIGHT1`, enables it in

Fig. 7.12 Artifact on the surface of the shape due to the degenerate normal

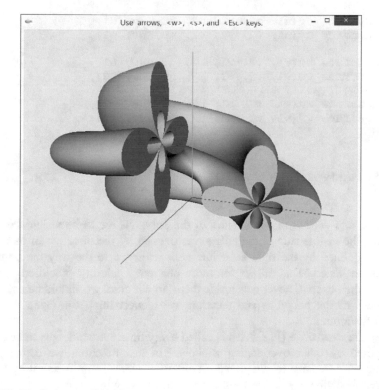

Fig. 7.13 Displaying a solid object

> Computation of coordinates x y
> and z of the solid as functions of
> parameters u, v and w

```
void Function(float u, float v, float w)
{
# define M_PI 3.14159265358979323846

x = ((0.35+w*0.5)*SIZE*sin(4 * u*M_PI - 0.5*M_PI)*cos(2 *
u*M_PI) + SIZE)*sin(v * 3 * M_PI / 2 + M_PI / 2);
y = (0.35 + w * 0.5)*SIZE*sin(4 * u*M_PI - 0.5*M_PI)*sin(2 *
u*M_PI) + SIZE * v;
z = ((0.35 + w * 0.5)*SIZE*sin(4 * u*M_PI - 0.5*M_PI)*cos(2 *
u*M_PI) + SIZE)*cos(v * 3 * M_PI / 2 + M_PI / 2);

}
```

Fig. 7.14 Computing coordinates of the solid object

```
void Solid(float u1, float u2, float v1, float v2, float w1,
float w2, int nu, int nv, int nw)
{
      Surface_uv(u1, u2, v1, v2, nu, nv, w1);
      Surface_uv(u1, u2, v1, v2, nu, nv, w2);

      Surface_vw(v1, v2, w1, w2, nv, nw, u1);
      Surface_vw(v1, v2, w1, w2, nv, nw, u2);

      Surface_wu(w1, w2, u1, u2, nw, nu, v1);
      Surface_wu(w1, w2, u1, u2, nw, nu, v2);

}
```

Fig. 7.15 Function displaying the solid object

the scene, and defines that both sides of the polygons are to be visible. Note that
normally the visible side of the polygon is defined by the direction of its normal,
which is defined by the right-hand rule with reference to the way the polygon's
vertices are listed. This allows for more efficient rendering of closed surfaces,
provided the observer never gets inside them. In our case, we should be able to see
both sides of the polygons and therefore we request this from OpenGL with a
special function.

Function reshape (Fig. 7.19) is called every time the window is to be resized
or exposed by other overlapping windows. In this function, we define a new
viewport with the size of the graphics window passed to this function by GLUT.
We need to define a new viewport if the size of the window changed. Then, we
prepare the modeling pipeline by first calling the projection transformation followed

```
void Surface_uv(float u1, float u2, float v1, float v2, int nu,
int nv, float w)
{
    float u, v, du,dv;
    int i, j;

    if (RendMode==1){glPolygonMode(GL_FRONT_AND_BACK,GL_LINE);
                     glColor4f(1.0, 1.0, 1.0, 1.0);}
    else {glEnable(GL_LIGHTING);
          glPolygonMode(GL_FRONT_AND_BACK,GL_FILL);}

    du = (u2 - u1) / (float)nu;
    dv = (v2 - v1) / (float)nv;

  for (j = 0; j < nv; j++) {
    v = v1+(v2-v1)*(float)j / (float)nv;
    glBegin(GL_TRIANGLE_STRIP);
    for (i = 0; i <=nu; i++) {
             u = u1+(u2-u1)*(float)i / (float)nu;
             if (i == 0) {
             Function(u, v+dv,w);
             Normal_uv(u, v+dv,w, du,dv,v2);
             glNormal3f(xn, yn, zn);
             glVertex3f(x, y, z);

.  .  .  .  .  .  .  .  .  .  .  .  .  .  .  .  .  .
}
```

Fig. 7.16 Function displaying the surface defined by *u* and *v* with fixed *w*

by setting the modeling transformations mode and loading the identity transformation into the stack of the affine transformations. We define the central projection with the near-clipping plane located at $3*SIZE$ from the camera, the far-clipping plane located at $40*SIZE$ from the camera, and the size of the window, through which we look at the scene, between $-2*SIZE$ and $2*SIZE$ (i.e. $4*SIZE \times 4*SIZE$).

In function draw (Fig. 7.20), first of all we clear the background with the background color $r = 0$, $g = 0$, $b = 0.2$. Then, we clear all the buffers to prepare for new rendering. Then, we define affine transformations in the reverse order (as for the column-represented position vector). Therefore, the shape, which is to be defined in function Curve, Surface_uv or Solid, is first rotated about the vertical axis Y, then about the horizontal axis X, and finally translated back by $-5*SIZE$ along the negative axis Z to fit into the viewing volume. The last transformation is needed because in OpenGL the camera is defined at the origin looking at the negative Z direction. Last function called in draw is glutSwapBuffers. It swaps the visible and the *auxiliary* buffers displayed on the screen with the new image just created.

```
void Surface_vw(float v1, float v2, float w1, float w2, int nv,
int nw, float u)
{
. . . . . . . . . . . . . . . . . .
}


void Surface_wu(float w1, float w2, float u1, float u2, int nw,
int nu, float v)
{
. . . . . . . . . . . . . . . . . .
}


void Normal_uv(float u, float v, float w, float du, float dv,
float v2)
{
. . . . . . . . . . . . . . . . . .
}


void Normal_vw(float u, float v, float w, float dv, float dw,
float w2)
{
. . . . . . . . . . . . . . . . . .
}


void Normal_wu(float u, float v, float w, float dw, float du,
float u2)
{
. . . . . . . . . . . . . . . . . .
}
```

Fig. 7.17 Profiles of two functions displaying the surfaces defined by v and w with fixed u, and by w and u with fixed v as well as the three functions calculating the respective normal

The interactive part of the program is defined in functions hotkey and sfunc. Function hotkey (Fig. 7.21) switches the rendering modes by setting variable RendMode and selecting GL_SMOOTH shading modes. When the "Esc" key is pressed, the execution of the program will be ended.

Function sfunc increases or decreases the rotation angle when the arrow keys are hit (Fig. 7.22).

Finally, we have main program which consists of only GLUT functions. These functions initiate OpenGL, define its modes, open the graphics window, specify all the callback functions which we have considered above, and start the event loop idle (Fig. 7.23).

```
#include <stdio.h>
#include <GL/glut.h>

#define   SIZE 1.0 /* a unit size used in the application */
static float x, y, z, xn, yn, zn; /*coordinates of the vertices
and normal */

int RendMode; /* May take values 1 or 2: for wireframe and polygon
modes. */
double alpha, beta;   /* Angles of rotation about vertical and
horizontal axes. */

// Defining a point light source parameters
GLfloat light_position1[] = {    0.0,    0.0, 100.0, 1.0 };
GLfloat light_ambient[]    = {    0.1,    0.1,   0.1, 1.0 };
GLfloat light_diffuse[]    = {    1.0,    1.0,   1.0, 1.0 };
GLfloat light_specular[]   = {    1.0,    1.0,   1.0, 1.0 };

//=============================================================
static void Init(void)
{
     alpha=-20.0; beta=20.0; RendMode=1;

// Setting up a point light source
     glLightfv(GL_LIGHT1, GL_AMBIENT,  light_ambient);
     glLightfv(GL_LIGHT1, GL_DIFFUSE,  light_diffuse);
     glLightfv(GL_LIGHT1, GL_SPECULAR, light_specular);
     glLightfv(GL_LIGHT1, GL_POSITION, light_position1);

// Enabling lighting with the light source #1
     glEnable(GL_LIGHTING);
     glEnable(GL_LIGHT1);

// Enabling both side illumination for the polygons and hidden
surface/line removal
     glLightModeli(GL_LIGHT_MODEL_TWO_SIDE, GL_TRUE);
     glEnable(GL_DEPTH_TEST);
}
```

Fig. 7.18 Functions Init

7.2.4 Animation and Surface Morphing with OpenGL and GLUT

In this section, we will implement animation of surface morphing as it was presented in Chap. 4. This is done in the book software repository project *OpengGL/Morphing*. In the software code *morphing.c*, morphing transformations between 25 different parametrically defined surfaces (Table 7.1) is defined as a continuous animation. The individual surfaces are defined and displayed in a way similar to how it was defined

```
static void reshape( int width, int height )
{
    if (width < height) { height = width; }
    else { width = height; };
    glutReshapeWindow(width, height);
    glViewport(0, 0, (GLint)width, (GLint)height);
    glMatrixMode (GL_PROJECTION);
    glLoadIdentity ();
    glFrustum(-1*SIZE, 1*SIZE, -1*SIZE, 1*SIZE, 2*SIZE, 40*SIZE);
    glMatrixMode(GL_MODELVIEW);
    glLoadIdentity ();
}
```

Fig. 7.19 Function reshape

```
static void draw( void )
{
        glClearColor(0.9, 0.9, 0.9, 0.0);
        glClear(GL_COLOR_BUFFER_BIT | GL_DEPTH_BUFFER_BIT);
        glPushMatrix();

        // Placement and rotation of the scene.
        glTranslatef(0.0, 0.0, -5 * SIZE);
        glRotatef(beta, 1.0, 0.0, 0.0);
        glRotatef(alpha, 0.0, 1.0, 0.0);
        CoordinateSystem();
        //Curve(0, 1, 100 );
        //Surface_uv(0,1,0,1,100,100,0);
        Solid(0, 1, 0, 1, 0, 1, 100, 100, 100);
        glFlush();
        glPopMatrix();
// swap animation buffers to display the current frame
        glutSwapBuffers();
}}
```

Fig. 7.20 Function draw

in Sect. 7.2.3 except that the normal vectors are calculated by differentiating the three parametric functions $x = f_1(u, v)$, $y = f_2(u, v)$, $z = f_3(u, v)$ by the variables u and v which will give us two vectors $\mathbf{n_1}$ and $\mathbf{n_2}$ tangent to the surface in the directions u and v. The cross product of these vectors yields the surface normal vector. Then, if one surface is defined with parametric formulas $x = f_1(u, v)$, $y = f_2(u, v)$, $z = f_3(u, v)$, and another one with parametric formulas $x = g_1(u, v)$, $y = g_2(u, v)$, $z = g_3(u, v)$, the morphing transformation can be defined as a linear transformation of coordinates:

```
static void hotkey(unsigned char k, int x, int y)
{
// Here we are processing keyboard events.
switch (k)
{
case 27:
     exit (0);
break;

// Switch to wireframe rendering
case 'w':
     RendMode=1;
break;

// Switch to smooth shading
case 's':
     RendMode=2;
     glShadeModel(GL_SMOOTH);
break;
}           }
```

Fig. 7.21 Function hotkey

$$x = f_1(u, v) + t(g_1(u, v) - f_1(u, v))$$

$$y = f_2(u, v) + t(g_2(u, v) - f_2(u, v))$$

$$z = f_3(u, v) + t(g_3(u, v) - f_3(u, v))$$

$$t \in [0, 1]$$

Now, the variable surface changing subsequently through all the 25 surfaces will be displayed. The variable t will control the animation frame to be displayed. Swing animation can be implemented by incrementing t from 0 to 1 followed by decrementing from 1 back to 0, and so on.

After you run this program, right click at the window to change the options. Note that all the formulas have the same parameter range to make the polygonization easier to implement. In reality, the parameters of two surfaces and their ranges can be different, and you will need first to modify the formulas so that they will have the same parameters and be in the same range.

```
static void sfunc(int k, int x, int y)
{
// Here we can process function keys and other special key
events
   switch (k)
   {

// Rotate to the left
      case GLUT_KEY_LEFT:
      alpha-=3.0;
      break;

// Rotate to the right
      case GLUT_KEY_RIGHT:
      alpha+=3.0;
      break;

// Rotate to the left
      case GLUT_KEY_UP:
         beta -= 3.0;
         break;

// Rotate to the right
      case GLUT_KEY_DOWN:
         beta += 3.0;
         break;
}
}
```

Fig. 7.22 Function `sfunc`

7.2.5 Interactive Solid Modeling with OpenGL and GLUT

In this section, we will write a program which will create a solid shape which is initially defined by a simple implicit function and then interactively modified by other functions which will add and remove solid materials to and from the shape. We will also be able to rotate the shape, change the parameters of the interactive operations, change the size of the graphics window, overlay it by other windows and iconify/de-iconify it. This is done in the book software repository project *OpengGL/Modeling*.

As a basic shape, we will use an origin-centered sphere with radius $r = 0.8$ defined by

$$r^2 - x^2 - y^2 - z^2 \geq 0$$

We will interactively modify this sphere by two types of operations: adding other spheres to the shape and removing other spheres from the shape. These

```
static void idle( void )
{
/* This function will call draw() as frequent as possible thus
enabling us to make interaction and animation */

    draw();
}


void main( int argc, char *argv[] )
{

    glutInit(&argc, argv);
    glutInitDisplayMode (GLUT_DOUBLE | GLUT_RGB );
    glutInitWindowSize (650, 650);
    glutInitWindowPosition (100, 100);
    glutCreateWindow ("Use  arrows,  <w>,  <s>, and  <Esc>
keys.");

    Init();
    glutReshapeFunc(reshape);
    glutIdleFunc(idle);
    glutDisplayFunc(draw);
    glutKeyboardFunc(hotkey);
    glutSpecialFunc(sfunc);

// The main event loop is started here.
    glutMainLoop();
}
```

Fig. 7.23 Functions idle and main

operations will simulate clay modeling. This depositing and removing of material will be implemented with two types of set-theoretic operations: *union and subtraction*, and *union and subtraction with blending*.

Union and subtraction are defined as in Chap. 2:

$$union(x, y, z) = \max(f_1(x, y, z), f_2(x, y, z)) \geq 0$$

$$subtr(x, y, z) = \min(f_1(x, y, z), -f_2(x, y, z)) \geq 0$$

Union and subtraction with blending are defined as follows:

$$union_{blend}(x, y, z) = f_1(x, y, z) + f_2(x, y, z) + \sqrt{f_1(x, y, z)^2 + f_2(x, y, z)^2}$$
$$+ \frac{p_1}{1 + \left(\frac{f_1(x,y,z)}{p_2}\right)^2 + \left(\frac{f_2(x,y,z)}{p_3}\right)^2}$$

Table 7.1 Parametric formulas of the surfaces are displayed in Fig. 7.24. In each formula φ and θ are in the range from 0 to 2π

1	$x = 1.6\,(\cos(0.5\varphi))^3$ $y = 1.6\,(\cos\theta\sin(0.5\varphi))^3$ $z = 1.6\sin\theta\sin(0.5\varphi)$	13	$x = 0.5\,(\cos\varphi + 0.5\cos(2\varphi))$ $y = 0.5\cos\theta\,(2 + \sin\varphi - 0.5\sin(2\varphi))$ $z = 0.5\sin\theta\,(2 + \sin\varphi - 0.5\sin(2\varphi))$
2	$x = 1.5\varphi\cos(\theta)/2\pi$ $y = 1.5\varphi\sin\theta\cos(\theta)/2\pi$ $z = 1.5\varphi\,(\sin\varphi)^5/2\pi$	14	$x = 0.1\,(3\cos(\varphi) + 0.8\cos(3\varphi))$ $y = 0.2\cos(\theta)(5 + 3\sin(\varphi) - 0.8\sin(3\varphi))$ $z = 0.2\sin(\theta)(5 + 3\sin(\varphi) - 0.8\sin(3\varphi))$
3	$x = \varphi\cos(\theta)/2\pi$ $y = \varphi\sin(\theta)/2\pi$ $z = \varphi\,(\sin\varphi\sin\theta)/2\pi$	15	$x = 0.15\,(4\cos(\varphi) - \cos(4\varphi))$ $y = 0.15\cos(\theta)(6 + 4\sin(\varphi) - \sin(4\varphi))$ $z = 0.15\sin(\theta)(6 + 4\sin(\varphi) - \sin(4\varphi))$
4	$x = \cos(0.5\varphi)\sin(0.5\varphi)$ $y = \cos\theta\sin(0.5\varphi)$ $z = \sin\theta\sin(0.5\varphi)$	16	$x = \cos(2\pi\varphi\sin(0.5\varphi))$ $y = \cos\theta\sin(0.5\varphi)$ $z = \sin\theta\sin(0.5\varphi)$
5	$x = \cos(0.5\varphi)$ $y = \cos\theta\sin(0.5\varphi)$ $z = \sin\theta\cos(0.5\varphi)$	17	$x = \sin\theta$ $y = \cos\varphi$ $z = \sin\theta\sin\varphi$
6	$x = (1 + 0.25\cos\theta)\cos\varphi$ $y = (1 + 0.25\cos\theta)\sin\varphi$ $z = 0.25\sin\theta$	18	$x = (\cos\theta + 1)\cos\varphi$ $y = \sin\theta\cos\varphi$ $z = \sin\varphi$
7	$x = \cos\theta\,(\sin(0.5\varphi))^3$ $y = \sin\theta\,(\sin(0.5\varphi))^3$ $z = \cos(0.5\varphi)$	19	$x = 0.5\,(\cos\varphi + 2)\cos\theta$ $y = 0.5\,(\sin\varphi + 2\theta/\pi - 2)$ $z = -0.5\,(\cos\varphi + 2)\sin\theta$

(continued)

Table 7.1 (continued)

8	$x = 0.25\varphi \cos\theta$ $y = 0.25\theta - 0.5$ $z = 0.25\varphi \sin\theta$		20	$x = 0.2\theta \sin\varphi$ $y = 0.2\varphi \sin\theta$ $z = 0.2\varphi \cos\theta$
9	$x = 0.5\cos\theta\left(-3 + \cos\varphi(1 + \cos\varphi)\right)$ $y = 0.5\sin\varphi\left(1 + \cos\varphi\right)$ $z = 0.5\sin\theta\left(-3 + \cos\varphi(1 + \cos\varphi)\right)$		21	$x = 1.5(\theta/\pi - 1)$ $y = 0.15\cos\left(5\sqrt{(2\theta/\pi - 2)^2 + (2\varphi/\pi - 2)^2}\right)$ $z = 1.5(\varphi/\pi - 1)$
10	$x = 0.5\left((\varphi - 0.5\sin\varphi) - 3\right)$ $y = 0.5\cos\theta\left(1 - 0.5\cos\varphi\right)$ $z = 0.5\sin\theta(1 - 0.5\cos\varphi)$		22	$x = 1.5\cos\theta\sin\varphi$ $y = 1.5\sin\theta\sin\varphi$ $z = 0.15\sqrt{\theta^2 + \varphi^2}\cos\varphi$
11	$x = 2.5\varphi / (1 + \varphi^3)$ $y = 2.5\cos\theta\,\varphi^2 / (1 + \varphi^3)$ $z = 2.5\sin\theta\,\varphi^2 / (1 + \varphi^3)$		23	$x = 2(\cos\theta)^3(\sin 0.5\varphi)^5$ $y = 2(\sin\theta)^3(\sin 0.5\varphi)^5$ $z = 2(\sin 0.5\varphi)^5\cos 0.5\varphi$
12	$x = 0.25\left(2\varphi - \sin(2\varphi) - 2\pi\right)$ $y = 0.5\cos\theta\left(2 - \cos(2\varphi)\right)$ $z = 0.5\sin\theta\left(2 - \cos(2\varphi)\right)$		24	$x = 2(\cos\theta)^3(\sin 0.5\varphi)^3$ $y = 2(\sin\theta)^3(\sin 0.5\varphi)^3$ $z = 2(\cos 0.5\varphi)^3$

$$subtr_{blend}(x, y, z) = f_1(x, y, z) + f_2(x, y, z)$$
$$- \sqrt{f_1(x, y, z)^2 + f_2(x, y, z)^2} - \frac{p_1}{1 + \left(\frac{f_1(x,y,z)}{p_2}\right)^2 + \left(\frac{f_2(x,y,z)}{p_3}\right)^2}$$

where p_1, p_2, and p_3 and parameters defining the amount of blending.

The program consists of three parts:

- Shape Modeling
- Shape Rendering
- User Interaction.

In Figs. 7.26 and 7.29, the source codes of the shape modeling part of the program are shown. The modeling is performed in the world coordinate system with coordinates x, y, z and within an origin-centered bounding box with size 2.2. We assume that the shape can be rotated when modifications are being made to it. Therefore, to accelerate the rendering, the *cos* and *sin* functions of the respective rotation angles are to be stored in the data structure by the part of the program implementing its user interface.

While modeling the shape, we form a binary constructive solid geometry tree which defines the shape. In the nodes of this tree, we will store the type of operation and its parameters, while in the leaves we will store parameters of the tools such as coordinates of the point where the interactive tool is applied, and the size of the tool. Each time when we need to visualize the shape, we will traverse this tree starting from the original shape and move further on through all the nodes thus evaluating the resulting function which defines the resulting shape.

The program can be further modified by adding functions of different original shapes as well as other functions of tools.

Function `ShapeFunction(x,y,z)` can now be used with any software capable of rendering shapes defined with functions $g(x, y, z) = f(x, y, z) \geq 0$.

We will render the shape using a simple ray tracing program, with the source code shown in Figs. 7.27 and 7.28.

The observer is located at infinity with the view vector $\boldsymbol{V} = [0\ 0\ 1]$. There is one light source defined by the light vector $\boldsymbol{L} = [0.577\ 0.577\ 0.577]$ and intensity 1.0 from the domain $[0.0\ 1.0]$. The object will be rendered in gray color on the blue background. Ambient coefficient is 0.1, diffuse and specular reflection coefficients are 0.8, and the specular reflection exponent is 120. Therefore, the illumination at each point where the ray intersects the surface is to be calculated according to the following formula:

$$I = \underbrace{0.1 \times 1.0}_{\text{Ambient light}} + \underbrace{0.8 \times 1.0 \times (\boldsymbol{N} \cdot \boldsymbol{L})}_{\text{Diffuse reflection}} + \underbrace{0.8 \times 1.0 \times (\boldsymbol{R} \cdot \boldsymbol{V})^{120}}_{\text{Specular reflection}}$$

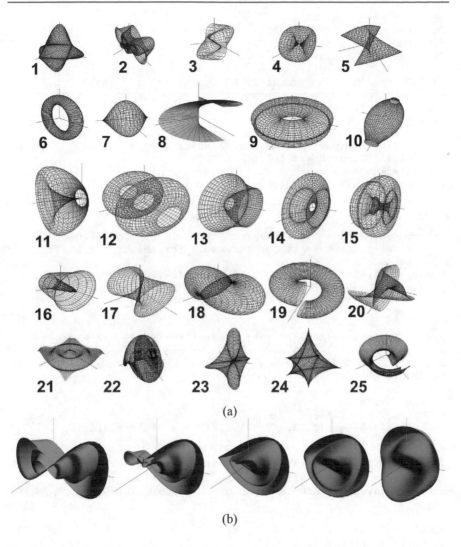

Fig. 7.24 a Surfaces of the shapes defined by formulas from Table 7.1; **b** Morphing animation between shapes 2 and 3

Since we do not know in advance what function will define our shape after each interactive operation, the intersection point of a ray with the shape will be calculated by incremental motion along the ray. The initial and final points on the ray are defined by the bounding box around the modeling area.

When moving the ray along a given incremental distance between the points, the function value is calculated for each point on the ray. When the sign of the function is negative, it means that we are still outside the shape and should continue moving along the ray. The change of the function's sign indicates that we are inside the

```
#include <stdio.h>
#include <math.h>
#include <stdlib.h>
#include <GL/glut.h>

// CSG tree storing parameters of operations and tools


typedef struct{
double xCoord, yCoord, zCoord, operationMode, blendMode,
     toolSize,param1,param2,param3,xoRtr,yoRtr,
     widthRtr,heightRtr,CC1,CC2,CS1,CS2;
} toolParameters;
toolParameters tool[1000];



int mode,operation_mode,nump,tool_type,width,height,flg=0,
     first=1,ico=0,xo_rtr,yo_rtr,width_rtr,height_rtr,width,
     height,IER;
double cos1,sin1,cos2,sin2,cr,cg,cb,angle1,angle2,rad,radInit,
     pa1,pa2,pa3,pa1m,pa2m,pa3m,delta,Zmax,Zmin,
     accuracy,xo_wcs,yo_wcs,scaleX,scaleY,width_wcs,height_wcs;


//*************************************************************
// Adding material with deformation of the work-piece
double    DepositBlend(double    f1,double    f2,double    a1,double
a2,double a3)
{
double fa1,fa2;
fa1=f1/a2; fa2=f2/a3;
return (f1+f2+sqrt(f1*f1+f2*f2)+a1/(1.+fa1*fa1+fa2*fa2));
}


//*************************************************************
// Removing material with deformation of the work-piece
double RemoveBlend(double f1,double f2,double a1,double a2,double
a3)
{
double fa1,fa2;
fa1=f1/a2; fa2=f2/a3;
return (f1+f2-sqrt(f1*f1+f2*f2)-a1/(1.+fa1*fa1+fa2*fa2));
}
```

Fig. 7.25 Source code of the data definition as well as two functions—DepositBlend, RemoveBlend—adding and removing material with blending

shape. After that, we will be approaching the surface of the shape by performing binary subdivision of the segment between the current and the previous points on the ray. The criterion for ending this subdivision is set by a certain function precision which is the difference between the current and the previous function values. Note that if the increment value for calculating coordinates of the point on the ray

```
// Function of the shape being modeled
double ShapeFunction(double X, double Y, double Z)
{
double
function_value,r,x,y,z,xp,yp,zp,xp1,yp1,zp1,tool_function,z0;
int i;
// Shape modeling in WCS XYZ within -1.1, +1.1
z0=-Y*sin2+Z*cos2;
y=Y*cos2+Z*sin2;
x=X*cos1-z0*sin1;
z=X*sin1+z0*cos1;
// Shape function "sphere". Any other function can be used here.
r=0.8;    function_value = r*r-x*x-y*y-z*z;
// Perform the operations
for(i=0;i<nump;i++)
{
// Translation to the origin
xp=x-tool[i].xCoord; yp=y-tool[i].yCoord; zp=z-tool[i].zCoord;
// Inverse rotation
xp1=xp*tool[i].CC1-zp*tool[i].CS1;       yp1=yp;
zp1=xp*tool[i].CS1+zp*tool[i].CC1;       xp=xp1;
yp=yp1*tool[i].CC2-zp1*tool[i].CS2;
zp=yp1*tool[i].CS2+zp1*tool[i].CC2;

// Tool function "sphere". Any other function can be used here.
tool_function=tool[i].toolSize*tool[i].toolSize-xp*xp-yp*yp-
zp*zp;

if(tool[i].operationMode==1){
// Removing material
    if(tool[i].blendMode==1)    function_value=min(function_value,-
tool_function );
        else               function_value=RemoveBlend(function_value,-
tool_function,
            tool[i].param1,tool[i].param2,tool[i].param3 );         }
else{
// Adding material
    if (tool[i].blendMode==1)
    function_value=max(function_value,tool_function );
        else
        function_value=DepositBlend(function_value,tool_function,
        tool[i].param1,tool[i].param2,tool[i].param3 );    }
}
   return function_value;
}
```

Fig. 7.26 Source code of function ShapeFunction(x,y,z) defining the shape created by modifying the original sphere with spherical tools

and the precision are large, we may miss some small parts of the shape and the resulting image will look coarse and distorted. On the other hand, if the increment and the precision are too small, the ray tracing will take more time.

```
void NormalCalculation(double x, double y, double z,
                       double *xn, double *yn, double *zn)
{ // Calculate normal to the surface at the point (x,y,z)
  double func, delta=0.00001;
  int degenerate=1;
  while(degenerate)
  { func = ShapeFunction(x, y, z);
    *xn = (ShapeFunction(x+delta, y, z)-func)/delta;
    *yn = (ShapeFunction(x, y+delta, z)-func)/delta;
    *zn = (ShapeFunction(x, y, z+delta)-func)/delta;
    if((*xn == 0.0) && (*yn == 0.0) && (*zn == 0.0))
      { delta=delta+delta; }
      else degenerate = 0;   }
  if(*zn < 0.0){*xn = -*xn; *yn = -*yn; *zn = -*zn;}
  delta=sqrt(*xn**xn+*yn**yn+*zn**zn);
  *xn=*xn/delta; *yn=*yn/delta; *zn=*zn/delta;
}
//***************************************************************
double Zcoordinate(int *miss,double X,double Y,double ZIncr,
double accuracy)
{ Calculate the intersection between the ray and the shape
  int ier, pro,flgdz=0;
  double Z,func,Zright,Zleft,Zcoord,Incr;
  Incr=ZIncr;  ier = 0;  Z = Zmax;
  func = ShapeFunction(X, Y, Z);    pro = 1;
  while(pro != 0)
    if(func < 0.0)
    {  Z = Z-Incr;
       if(Z < Zmin){ pro = 0;  ier = 1; }
       else func = ShapeFunction(X, Y, Z); }
    else
    {  if(flgdz){Z=Z+Incr;Incr=ZIncr/100.0;flgdz=1;}
       else{ Zright = Z; Zleft = Zright+Incr; pro = 0;}    }
  pro = 1;
  if(ier == 0)
  {  while(pro != 0)
     {Z = (Zright+Zleft)/2.0;
      func = ShapeFunction(X, Y, Z);
      if((fabs(func) > accuracy) && ((Zleft-Zright) > accuracy))
         if(func >= 0.0) Zright = Z;
         else Zleft = Z;
       else pro = 0;        }
     Zcoord = (Zright+Zleft)/2.0;    }
  *miss = ier;
  return Zcoord; }
```

Fig. 7.27 Source codes of the function NormalCalculation calculating normal to the surface and Zcoordinate computing the intersection point between the ray and the surface of the object

When the intersection point is thus calculated, we will compute the illumination at this point as a sum of the ambient, diffuse, and specular reflections. This will involve calculation of a normal to the surface at the intersection point, followed by computing

```
void Raytrace(void)
{
  float col[3900];
  int ic;
  double X,Y,Z,XN,YN,ZN,XL1,YL1,ZL1,NL,I,RV;
  int i, j;
   IER = 0;

// Light vector L
  XL1=0.577;   YL1=0.577;   ZL1=0.577;
// Accuracy of functon evaluation
  accuracy = 0.00001;

// scale factors from DCS to WCS
  scaleX = width_wcs/width; scaleY = height_wcs/height;

  for(i=yo_rtr; i<yo_rtr+height_rtr; i++)
  {
    ic=0;
    for(j=xo_rtr; j<xo_rtr+width_rtr; j++)
    { X = (scaleX * (double)j) + xo_wcs;
      Y = (scaleY * (double)(height - i)) + yo_wcs;
      Z = Zcoordinate(&IER,X,Y,0.1,accuracy);
      if(IER == 0)
      { NormalCalculation(X,Y,Z,&XN,&YN,&ZN);
// Diffuse reflection NL. If NL<0 - point is not visible from
// the source
        NL=XN*XL1+YN*YL1+ZN*ZL1;
        if(NL < 0.0) NL = 0.0;

// Specular reflection RV R=2(N*L)N-L       V=[0 0 1]
        RV=2*NL*ZN-ZL1;
               if(RV < 0.0) RV = 0.0;

// Illumination=ambient+diffuse+specular reflections
        I = 0.1 + 0.8*NL + 0.8*pow(RV,120);
        if(I > 1.0) I = 1.0;
        col[ic]=I*cr; col[ic+1]=I*cg; col[ic+2]=I*cb; }
      else {col[ic]=0.0; col[ic+1]=0.6; col[ic+2]=1.0; }
      ic=ic+3;
      }

glRasterPos2i(xo_rtr,height-i-1);
glDrawPixels(width_rtr,1,GL_RGB,GL_FLOAT,col);   }
}
```

Fig. 7.28 Source code of function Raytrace which ray traces the shape defined by ShapeFunction $(x,y,z) = f(x,y,z) \geq 0$

the dot products of normal N and light vector L as well as view V and reflected ray R vectors. The normal to the implicitly defined surface $f(x, y, z) = 0$ is to be calculated as the gradient of the function:

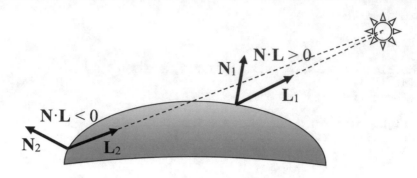

Fig. 7.29 Detecting illuminated and non-illuminated points

$$N = \left[\begin{array}{ccc} \frac{\partial f(x,y,z)}{\partial x} & \frac{\partial f(x,y,z)}{\partial y} & \frac{\partial f(x,y,z)}{\partial z} \end{array} \right]$$

Points where the angle between the normal to the surface **N** and the light vector **L** are greater than 90° (**N·L** < 0) are not illuminated by the light source (Fig. 7.29) and their contribution to the diffuse reflection is 0. Similarly, when calculating specular reflections, negative **R·V** values will be ignored.

To achieve interactivity, each time we will ray trace only a small part of the image enclosing the area where the shape modification is made. The size of this window is proportional to the current tool size. The parameters of these windows updating the resulting shape will be also stored in the model's data structure. The whole image will be ray traced only when the shape rotates, the image size changes, or if it is requested by the user. Depending on the number of operations, this may take a much longer time than ray tracing of a small part of the image.

The source code of the remaining third part of the program, where all the interactive operations are processed, is shown in Figs. 7.30, 7.31, 7.32, 7.33 and 7.34.

Function `Init` defines the CSG tree data structure as well as all the required variables.

Function `Draw` will be called each time when the whole graphics window or a part of it is to be updated. It calls the ray tracing function `Raytrace` as well as copies the content of the `FRONT` buffer to the auxiliary `BACK` buffer to keep a copy of the image in the case if the graphics window is overlapped with another window which will require quick redrawing of the affected image.

Function `Reshape` is called each time the graphics window is resized or exposed. It sets the new window parameters and prepares for redrawing in the window.

Function *mouse_press* is called when the left mouse button is clicked. This function computes 3D coordinates of the point selected on the surface of the shape as well as writes them and all the tool's parameters to the data structure.

Function `mykey` changes the size of the tool proportionally to integer numbers 1–9 when keys "1"–"9" are hit.

```
static void Init()
{
// Set initial parameters
mode=1;      angle1=0.0;      angle2=0;      rad=0.05;      radInit=0.05;
operation_mode=1;    nump=0;    tool_type=1;    flg=0,first=1,ico=0;
pa1m=pa1=0.0025;  pa2m=pa2=0.01;  pa3m=pa3=0.01;  cos1=1;  sin1=0;
cos2=1;      sin2=0;      xo_rtr=0;      yo_rtr=0;      width_rtr=width=0;
height_rtr=height=0;    delta=3.14159265/36.0;      cr=1.0;  cg=1.0;
cb=1.0;      Zmax  =  2.0;  Zmin  =  -1.0;  xo_wcs=-1.1;  yo_wcs=-1.1;
width_wcs=2.2; height_wcs=2.2;
}



//*****************************************************************
static void Draw(void)
{
// Draw the whole or partial image
if(first!=1 && ico==1)return;
if(flg==0){
     glMatrixMode(GL_PROJECTION);
     glLoadIdentity();
     gluOrtho2D(0.0,width,0.0,height);
     Raytrace(); flg=1; first=0;}
glutSwapBuffers();
glReadBuffer(GL_FRONT);
glDrawBuffer(GL_BACK);
glRasterPos2i(0,0);
glCopyPixels(0,0,width,height,GL_COLOR);
}



//*****************************************************************
static void Reshape(int w, int h)
{
// Change the size of the image
if(w==0 && h==0){ico=1; return;}
if(ico){ico=0;}
else{width=w; height=h;}
xo_rtr=0; yo_rtr=0; width_rtr=width; height_rtr=height;
glViewport(0, 0, (GLint)w, (GLint)h);
flg=0;
}
```

Fig. 7.30 Functions Init, Draw, and Reshape

Function sfunc changes the rotation angles of the shape about the vertical and the horizontal axes when keys ← , → , ↑, ↓ are hit.

Function mymenu processes the events which are generated when the pop-up menu commands are selected. These commands switch the operations modes between Remove and Deposit material, switches on and off blending for

```
static void mouse_press(int button, int state, int x, int y)
{ int width_rtr1,height_rtr1;
double k,X,Y,Z,XPN,YPN,ZPN,XN,YN,ZN,hc1,hc2;
// The following is involved when the left button pressed
if((button == GLUT_LEFT_BUTTON) && (state == GLUT_DOWN)){
// Derive 3D coordinates of the tool in the WCS
X = (scaleX * (double)x) + xo_wcs;
Y = (scaleY * (double)(height - y)) + yo_wcs;
Z = Zcoordinate(&IER,X,Y,0.1,accuracy);
NormalCalculation(X,Y,Z,&XN,&YN,&ZN);
tool[nump].yCoord=Y*cos2+Z*sin2;              Z=-Y*sin2+Z*cos2;
tool[nump].xCoord=X*cos1-Z*sin1;
tool[nump].zCoord=X*sin1+Z*cos1;
// Calculate and store the tool parameters
YPN=YN*cos2+ZN*sin2;        ZN=-YN*sin2+ZN*cos2;
XPN=XN*cos1-ZN*sin1;
ZPN=XN*sin1+ZN*cos1;
hc1=XPN*XPN+ZPN*ZPN;
hc2=sqrt(hc1+YPN*YPN);
hc1=sqrt(hc1);
tool[nump].CC1=ZPN/hc1;
tool[nump].CS1=XPN/hc1;
tool[nump].CC2=hc1/hc2;
tool[nump].CS2=YPN/hc2;
tool[nump].operationMode=operation_mode;
tool[nump].blendMode=mode;
tool[nump].toolSize=rad;
tool[nump].param1=pa1;
tool[nump].param2=pa2;
tool[nump].param3=pa3;
// Calculate the parameters of the rendering window for the tool
if(mode==1)k=1.2; else k=1.6;
width_rtr1=(int)(k*(width-1)*(rad/width_wcs));
height_rtr1=(int)(k*(height-1)*(rad/height_wcs));
xo_rtr=x-width_rtr1; if(xo_rtr<0)xo_rtr=0;
yo_rtr=y-height_rtr1; if(yo_rtr<0)yo_rtr=0;
width_rtr=width_rtr1+width_rtr1;
if(width_rtr>width-1-xo_rtr)width_rtr=width-1-xo_rtr;
height_rtr=height_rtr1+height_rtr1;
if(height_rtr>height-1-yo_rtr)height_rtr=height-1-yo_rtr;
tool[nump].xoRtr=xo_rtr;
tool[nump].yoRtr=yo_rtr;
tool[nump].widthRtr=width_rtr;
tool[nump].heightRtr=height_rtr;
nump++;     flg=0;
Draw();           }
}
```

Fig. 7.31 Function mouse_press

these operations, undo the results of operations as many times as needed, and
refresh the whole image by redrawing it if some artifacts are created during
partial updating of the image.

```
static void mykey(unsigned char k, int x, int y)
{ // Process the event when a hot-key command is typed
   int size;
   switch (k) {
      case 27: // Exit program
      exit (0);
      break;
      case '1':
      case '2':
      case '3':
      case '4':
      case '5':
      case '6':
      case '7':
      case '8':
      case '9':
// Tool size changes
// ascii codes for "1" - "9" are 49 to 57
         size=k-48;
         rad=radInit*size;
         pa1=pa1m*size; pa2=pa2m*size; pa3=pa3m*size;
      break;
   }
flg=1; Draw();
}

static void sfunc(int k, int x, int y)
{ // Process the event when arrow keys are pressed
   switch (k) {
     case GLUT_KEY_UP:     // Rotate up
     angle2=angle2-delta; cos2=cos(angle2); sin2=sin(angle2);
     break;

     case GLUT_KEY_DOWN:   // Rotate down
     angle2=angle2+delta; cos2=cos(angle2); sin2=sin(angle2);
     break;

     case GLUT_KEY_LEFT:   // Rotate left
     angle1=angle1-delta; cos1=cos(angle1); sin1=sin(angle1);
     break;

     case GLUT_KEY_RIGHT: // Rotate right
     angle1=angle1+delta; cos1=cos(angle1); sin1=sin(angle1);
     break;          }
   xo_rtr=0; yo_rtr=0; width_rtr=width; height_rtr=height;
   flg=0; Draw();
}
```

Fig. 7.32 Functions mykey and sfunc

```
static void mymenu(int k)
{ // Process the events when pop-up menu items are selected
     switch(k)
     {

     case 1:
// Remove material
          operation_mode=1;
          break;
     case 2:
// Add material
          operation_mode=2;
          break;
     case 3:
// Undo
          if(nump){
          nump--;
          xo_rtr=tool[nump].xoRtr;
          yo_rtr=tool[nump].yoRtr;
          width_rtr=tool[nump].widthRtr;
          height_rtr=tool[nump].heightRtr;
          flg=0;
          Draw();}
          break;
     case 4:
// Blend on
          mode=2;
          break;
     case 5:
// Blend off
          mode=1;
          break;
     case 7:
// New
          nump=0;
     case 6:
// Redraw the whole image
          xo_rtr=0; yo_rtr=0;
          width_rtr=width;
          height_rtr=height;
          flg=0;
          Draw();
          break;

     }
     flg=1;
}
```

Fig. 7.33 Function mymenu

```
void main( int argc, char *argv[] )
{ // Main program using GLUT
// Initialize everything
   glutInit(&argc, argv);
// Set OpenGL modes
   glutInitDisplayMode (GLUT_DOUBLE | GLUT_RGB );
// Define the window size
   glutInitWindowSize (300, 300);
// Position the window
   glutInitWindowPosition (100, 100);
// Create graphics window
   glutCreateWindow ("Shape Modeler");

// Function to call first to set the initial values
   Init();
// Function when the size changes
   glutReshapeFunc(Reshape);
// Function to draw the image
   glutDisplayFunc(Draw);

// Function for keys
   glutKeyboardFunc(mykey);

// Function for special keys
   glutSpecialFunc(sfunc);

// Setting up pop-up menu items
   glutCreateMenu(mymenu);
          glutAddMenuEntry("Remove", 1);
          glutAddMenuEntry("Deposit", 2);
          glutAddMenuEntry("Undo", 3);
          glutAddMenuEntry("Blending On", 4);
          glutAddMenuEntry("Blending Off", 5);
          glutAddMenuEntry("Refresh", 6);
          glutAddMenuEntry("New", 7);
// Menu on right button pressed
   glutAttachMenu (GLUT_RIGHT_BUTTON);

// Function to call for mouse events
   glutMouseFunc(mouse_press);

// Initialize the event loop
   glutMainLoop();
}
```

Fig. 7.34 Function main

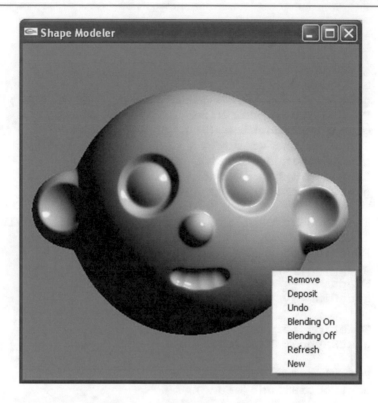

Fig. 7.35 Snapshot of the modeling session with the interactive program

Finally, in the main function, we initiate GLUT and OpenGL, define parameters of the graphics window and all the callback functions for the events when the window is resized or redrawn, and keys or mouse buttons are pressed.

In Fig. 7.35, a snapshot of the program's window is shown. The shape displayed there was created with 14 spheres unified and subtracted with blending.

You may experiment with different basic shape functions using them as tools and original shapes. You may also notice that when more functions are added to the model, the redrawing of the whole shape will become slower. Some optimizations to the data model will be required then.

7.3　Drawing with POV-Ray

7.3.1　The Persistence of Vision Ray Tracer

When personal computers were invented, computer image generation was only a dream. Computer graphics programs were usually power-hungry applications requiring lots of memory and speed. Now, computer image generation is everywhere, from advertising and movies to games and art. The incredible growth in the popularity of computer-generated art has developed into a quest for the ultimate goal: a perfect photorealistic image.

The most straightforward method of creating stunning computer graphics images is by using ray tracing. We discussed the basics of ray tracing in Sect. 3.4.1. Let's try it now. We will be learning ray tracing using *The Persistence of Vision Ray Tracer* (POV-Ray) [7]. POV-Ray is a copyrighted freeware program that lets a user easily create fantastic, three-dimensional, photorealistic images on just about any computer. POV-Ray reads standard ASCII text files that describe the shapes, colors, textures, and lighting in a scene and mathematically simulates the rays of light moving through the scene to produce a photorealistic image.

No traditional artistic or programming skills are required to use POV-Ray. This is the best way to learn computer graphics even for those who have no technical background. First, you describe a picture in POV-Ray's scene description language (declarative style of programming), and then POV-Ray takes your description and automatically creates an image from it with near-perfect shading, perspective, reflections, and lighting.

POV-Ray offers a large set of 3D primitives (basic and advanced shapes), affine transformations (translation, rotation, and scaling), set-theoretic or Boolean operations, multiple light sources, ability to render shadows, rendering with antialiasing,[1] and different photorealistic techniques including textures. The standard POV-Ray package also includes a collection of sample scene files that illustrate the program's features. They can also be modified to create new scenes.

Let us create the scene file for a simple picture.

Download and install POV-Ray from its web page https://www.povray.org. **Then, run POV-Ray by clicking on its icon. The POV-Ray user interface will look as it is displayed in** Fig. 7.36. Explore its features. Render different scenes by selecting them from the *File/Open File...* menu and by clicking on the *Run* button. Select different sizes for the images as well as render them with and without antialiasing (option AA 0.3 and NO AA).

Now, it is time to create your first POf-Ray image.

In *File* menu, select *New File*. It will make a new empty page where you can type your scene description code.

In every computer graphics problem, we have to define the observer, the light sources, and the objects which are to be visualized. Let's first define the observer's position and viewing direction as shown in Fig. 7.37.

We defined the observer at the point with coordinates $(0, 2, -3)$. The observer is looking at the point with coordinates $(0, 1, 2)$. Note that a left-handed coordinate

---

[1]Antialiasing is a technique for smoothing jagged edges by blending shades of color, or gray along the edges. Images rendered with antialiasing have better quality, though they require longer time for ray tracing. In one antialiasing technique, blurred pixels are introduced by filtering the image, or individual elements, to remove spatial frequencies that are greater than the pixel sample rate by convolution. If high frequencies remain they may cause other visual artifacts such as Moiré patterns. An alternative and often preferable technique when ray tracing images is supersampling, where many samples per pixel are estimated and combined.

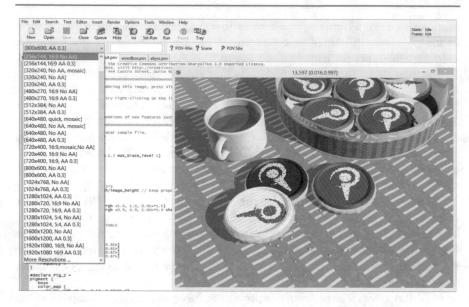

Fig. 7.36 POV-Ray user interface and the ray traced image

```
#include "colors.inc"
#include "shapes.inc"
#include "textures.inc"
camera {
        location  <0, 2, -3>
        up <0, 1, 0>
        right <4/3, 0, 0>
        look_at <0, 1,  2>
            }
```

Fig. 7.37 Defining the observer's position and the viewing direction

system is used here—this is the traditional way of defining coordinates in ray tracing software. Next, we will add a white light source at point $(2, 4, -3)$, and also define an object—a green sphere with center at point $(0, 1, 2)$ and radius 2 (Fig. 7.38). After that we only need to save the file and click at the *Run* button to ray trace the scene.

The sphere looks dull since only diffused reflection is used for its rendering. To add a highlight onto its surface, we can apply the Phong illumination model as it is shown in Fig. 7.39.

```
.  .  .  .  .  .  .  .  .  .  .  .  .  .

light_source { <2, 4, -3>
color White}

sphere { <0, 1, 2>, 2
      texture  {
      pigment {color Yellow}
   }
         }
```

Fig. 7.38 Defining the light source and a shape

```
sphere{ <0,1,2>, 2

texture {
      pigment {color Green}
      finish {phong 1}
      }
            }
```

Fig. 7.39 Adding a highlight on the surface

Next, we will make an appearance of a bumpy surface by defining so-called
normal perturbation. The surface will look like it has bumps on it. However, it is
just an illusion since we simply intervened into the rendering process and artificially
changed the directions of the normals. As a result, in place of a smooth shading of
the spherical surface, we have patterns as if the surface were bumpy (Fig. 7.40).

We can also define our own pattern on the surface by creating a *color map* and
setting a turbulence which will distort the basic pattern before it is applied to the
surface of the sphere. We will also define a background color to make it lighter
(Fig. 7.41).

There are many texture patterns predefined in the *textures.inc* file which was
included at the beginning of the code. We can use these textures by defining their
names and setting variable parameters (Fig. 7.42).

There are other basic objects which can be used in POV-Ray. In Fig. 7.43, we
put together in the same scene *Plane, Box, Cone,* and *Cylinder.* Plane and Box are
defined so that they are initially parallel to one of the three principal coordinate
planes. However, they can be then transformed with affine transformation. Cone
and Cylinder can be defined with open and closed-end caps.

```
sphere{ <0,1,2>, 2

texture{
pigment {color Green}

normal {bumps 0.4 scale 0.2}
finish {phong 1}
          }
                }
```

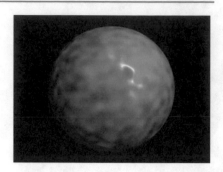

Fig. 7.40 Adding bumpiness

```
sphere{ <0,1,2>, 2
texture{pigment{ wood
color_map{
[0.0 color VeryDarkBrown]
[0.8 color DarkBrown]
[1.0 color DarkTan]}
turbulence 0.05
scale<0.6, 0.4, 1>}
finish {phong 1} }}
   background {
   color rgb <0.5, 0.5, 0.0> }
```

Fig. 7.41 Making a texture pattern

```
sphere{ <0,1,2>, 2

texture{
   pigment {
     Brown_Agate
     scale 3}
      finish {Shiny}
      }
          }
   background {
   color rgb <0.5, 0.5, 0.0> }
```

Fig. 7.42 Applying a predefined texture

Besides these basic shapes, there are also other advanced shapes: *Torus (Donut), Hyperboloid, Paraboloid, Bezier Patch, Height Fields, Blobs, Quadric, Quartic, Smooth Triangles*, and *Isosurface*.

```
plane{ <0,1,0>,0  pigment
{checker color White color Blue}}
box{<-1, 0, 1>,   <1, 2.5 , -1>
pigment{ color Yellow}
rotate y*40 translate <0, 0, 2> }
cone{<-1.3, 0.0, 0.0>, 0.2
     <-1.3, 1.0, 0.0>, 0.55
open
pigment { color Green}
finish {Shiny}}
cylinder{
<1.4, 0.5, 0.0>,
<2.5, 0.5, 3.0>, 0.5
pigment {color Red}
finish {Shiny}}
```

Fig. 7.43 Plane, Box, Cone, and Cylinder

Set-theoretic (Boolean) operations can be used for creating complex CSG objects from primitive objects. POV-Ray has three set-theoretic operations: *union, intersection,* and *difference.* Union is a default operation which is always applied when objects are put in the same scene. In Fig. 7.44, a sphere is unified with a cylinder. Each of the shapes retains its original color.

In Fig. 7.45, the same shapes intersect, and the result of their intersection is a cylinder with rounded end caps. The caps and the cylinder inherit colors of the respective shapes.

```
camera {
location   <2, 2, -3.5>
up <0, 1, 0>
right <4/3, 0, 0>
look_at <0, 1,  2>}
light_source { <2, 4, -3>
   color White}
```

```
light_source { <2, 4, -3>
color White}

union{
          sphere {<0, 1, 2>, 2
texture { pigment {color Red}   finish {phong 1} }  }
          cylinder {<0,1,-1.2>,<0,1,4.2>, 0.5
texture{Gold_Metal} }
               }
background { color rgb <0.0, 0.8, 0.8> }
```

Fig. 7.44 Union of two objects

```
intersection{
sphere {<0, 1, 2>, 2
    texture {
pigment {color Red}
    finish {phong 1}
    }    }
cylinder {<0, 1, -1.2>,
<0, 1, 4.2>, 0.5
texture {Gold_Metal}
    } }
```

```
background { color rgb <0.0, 0.8, 0.8> }
```

Fig. 7.45 Intersection of two objects

In Fig. 7.46, the same cylinder is subtracted from the sphere, and the result of this operation is a sphere with a cylindrical hole. The sphere and the sides of the hole inherit colors of the respective shapes.

Finally, in Fig. 7.47, we make a complex object by unifying the shape defined in Fig. 7.45 with a vertical cylinder.

Besides this quick introduction, you can go through the POV-Ray tutorial exercises. Click F1—it will display the help menu. From the *Contents* menu, select *Beginning Tutorial* and follow the instructions there. To simplify scene creation, an interactive shareware wireframe modeler Moray [8] supporting POV-Ray can be used.

7.3.2 Function-Based Shape Modeling with POV-Ray

POV-Ray allows for defining 3D shapes with implicit and parametric formulas. It can be done with an *isosurface* object. Mathematically, if the shape is defined with function $f(x, y, z)$ its surface is defined by an isosurface function $g = f(x, y, z)$. All

```
difference{
sphere {<0, 1, 2>, 2
    texture { pigment
    {color Red}
    finish {phong 1}
    }
        }
cylinder {<0, 1, -1.2>,
<0, 1, 4.2>, 0.5
texture {Gold_Metal}
    }    }
```

```
background { color rgb <0.0, 0.8, 0.8> }
```

Fig. 7.46 Difference of two objects

```
union{
   cylinder {<0,-3,1>,
             <0,3,1>,0.5
   texture{White_Wood}}
   difference{
    sphere {<0, 1, 2>, 2
    texture { pigment
             {color Red}
    finish {phong 1}}}
    cylinder {<0,1,-1.2>,
              <0,1,4.2>,0.5
    texture{Gold_Metal}}
                       }
                }
background { color rgb <0.0, 0.8, 0.8> }
```

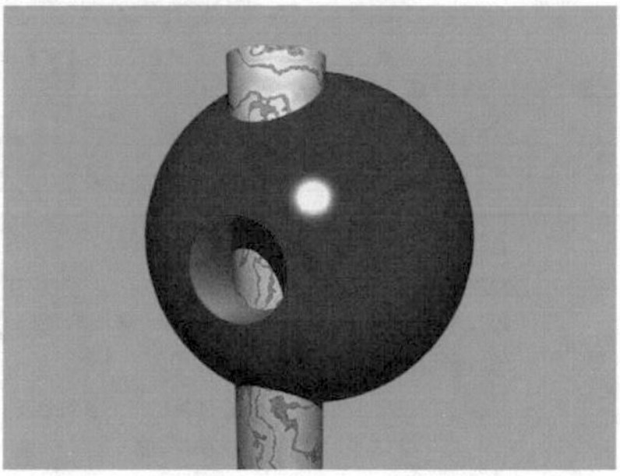

Fig. 7.47 Complex CSG object created with one sphere and two cylinders

points, which are tested against a defined function and equal a required threshold value, belong to the object's surface. It is obvious that POV-Ray couldn't test all the points in an infinite space, since it would take forever. To speed things up, points are sampled within a defined area and within a specified accuracy range. Then, the surface is created by interpolation between the matching points. This means that an isosurface is an approximation (accuracy depending on the settings) of the exact location of the function's surface. But for the vast majority of scenes, this is more than accurate enough.

In Chap. 3, we considered shape definition with functions $f(x, y, z) \geq 0$. In POV-Ray, the shapes are defined with functions $f(x, y, z) \leq 0$. To be consistent with this requirement of POV-Ray, we can simply place a minus sign in front of the formulas $f(x, y, z) \geq 0$ which were used in Chap. 3. Therefore, to ray trace a functionally defined sphere, we have to run the code listed in Fig. 7.48.

By increasing the exponent of $x, y,$ and z, we can make very different shapes which belong to the superellipsoid class (Fig. 7.49). To be able to display this shape properly, we have to increase the maximum gradient value—this is a major inconvenience of rendering isosurfaces since the rendering parameters must be tuned to the model to be displayed.

Further experiments with the same formula (originally an equation of a sphere) will make very different shapes (Fig. 7.50).

Next, we illustrate how the set-theoretic operations—union, intersection, and difference—can be incorporated into one formula (Fig. 7.51). In this example, we apply these operations to two spheres. The second sphere is translated by +1 in x-coordinate.

Let's now define a cube by intersecting 6 half-spaces bounded by planes, and then make a few experiments with this formula as well as understand how such half-spaces can be used for creating different polyhedrons (Fig. 7.52).

In the next exercise (Fig. 7.53), we first make a cylinder with radius 2 and height 2. This cylinder is Y-axis aligned. Next, we unify it with a sphere of radius 2 and

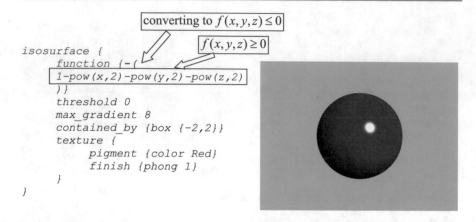

```
isosurface {
    function {-(
    1-pow(x,2)-pow(y,2)-pow(z,2)
    )}
    threshold 0
    max_gradient 8
    contained_by {box {-2,2}}
    texture {
        pigment {color Red}
        finish {phong 1}
    }
}
```

Fig. 7.48 Definition of a sphere in the isosurface object

```
isosurface {
function {-(
    1-pow(x,8)-pow(y,8)-pow(z,8)
    )}
    threshold 0
    max_gradient 256
    contained_by {box {-2,2}}
    texture {
        pigment {color Red}
        finish {phong 1}
    }
}
```

Fig. 7.49 Definition of a superellipsoid in the isosurface object

1-pow(x,2)-cos(y)*y-pow(z,2)   1-pow(x,2)-cos(y)*sin(y)-pow(z,2)

Fig. 7.50 Experimenting with a quadric function

(a)   `max(1-pow(x,2)-pow(y,2)-pow(z,2),`
       `      1-pow(x-1,2)-pow(y,2)-pow(z,2))`
(b)   `min(1-pow(x,2)-pow(y,2)-pow(z,2),`
       `      1-pow(x-1,2)-pow(y,2)-pow(z,2))`
(c)   `min(1-pow(x,2)-pow(y,2)-pow(z,2),`
       `      -(1-pow(x-1,2)-pow(y,2)-pow(z,2)))`

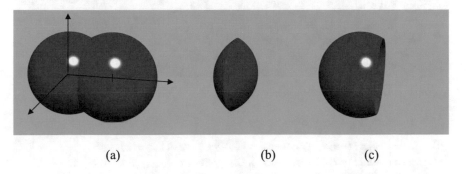

(a)   (b)   (c)

Fig. 7.51 Set-theoretic operations: **a** union, **b** intersection, and **c** difference of two spheres

center at point $(0, 0.5, 0)$. Finally, we'll scale the sphere up 10%. Note that all affine transformations are to be defined as inverse ones in the formulas.

With reference to Chap. 4, we can also illustrate how function-defined shape morphing works. In Fig. 7.54, we transform a sphere into a cube. The shape's function is a linear function of the sphere's functions and the cube's function.

Parametrically defined surfaces can be also rendered with POV-Ray. In Fig. 7.55, an example of such a parametric surface is given. This surface is defined as follows:

$$x = \frac{u \cos v}{2\pi}, \quad y = \frac{u \sin v}{2\pi}, \quad z = \frac{u \sin v \sin u}{2\pi}, \quad u, v \in [0, 2\pi]$$

7.4 Drawing with VRML/X3D

7.4.1 Introduction to VRML

Virtual Reality Modeling Language (VRML) and its current successor Extensible 3D (X3D) [9] are ISO[2] standard file formats and programing languages for describing interactive 3D objects and virtual worlds. They are designed to be used

---

[2]The International Organization for Standardization (ISO) is an international standard-setting body composed of representatives from various national standards organizations.

(a)   `min(1-x, 1-y, 1-z,   x+1, y+1, z+1)`
(b)   `min( min(1-x, 1-y, 1-z, x+1, y+1, z+1), x*y)`
(c)   `min( min(1-x, 1-y, 1-z, x+1, y+1, z+1), x+y)`

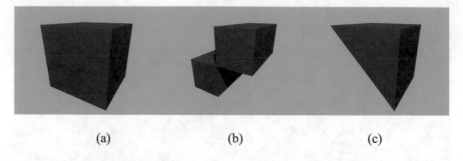

(a)   (b)   (c)

(d)   `min( min(1-x, 1-y, 1-z, x+1, y+1, z+1), x+z)`
(e)   `min( min(1-x, 1-y, 1-z, x+1, y+1, z+1), -x-z)`
(f)   `min( min(1-x, 1-y, 1-z, x+1, y+1, z+1), -x-y+z+1)`

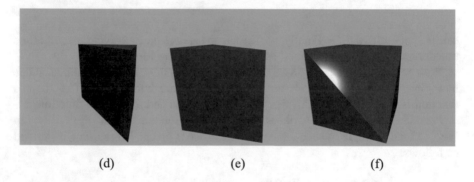

(d)   (e)   (f)

Fig. 7.52 Experimenting with a cube

on the internet, intranets, and local client systems. They are also intended to be universal interchange formats for integrated 3D graphics and multimedia. VRML and X3D are capable of representing static and animated dynamic 3D and multimedia objects with hyperlinks to other media such as text, sounds, movies, and images. VRML and X3D are following *declarative programming style* which tells the computer *what to do*. It differs from the *imperative programming style*, like in OpenGL, which tells the computer *how to do* things.

A VRML file is a collection of *objects* which are shapes, sounds, lights, and viewpoints. *Nodes* are fundamental building blocks of a VRML file. Each node can contain other nodes. Each node contains *fields* which hold the data for the node.

Let's consider how to create a simple VRML scene with one object: a shiny purple cylinder with a radius 3, and a height 6 on a blue background. The VRML

(a)   `min(1-pow(x,2)-pow(z,2),min(1-x,1-y,1-z,x+1,y+1,z+1))`

(b)   `max(1-pow(x,2)-pow(y-0.5,2)-pow(z,2),`
      `min(1-pow(x,2)-pow(z,2),`
      `min(1-x,1-y,1-z,x+1,y+1,z+1)))`

(c)   `max(1-pow(x/1.1,2)-pow(y/1.1-0.5,2)-pow(z/1.1,2),`
      `min(1-pow(x,2)-pow(z,2),`
      `min(1-x,1-y,1-z,x+1,y+1,z+1)))`

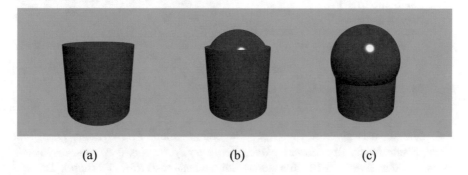

Fig. 7.53   Set-theoretic operations over a sphere and a cylinder

`0.2*(1-pow(x,2)-pow(y,2)-pow(z,2))`
`+0.8*min(1-x,1-y,1-z,x+1,y+1,z+1)`

Fig. 7.54   Morphing a sphere into a cube

```
parametric {
    function { u*cos(v)/(2*pi) }
    function { u*sin(v)/(2*pi) }
    function
    { u*sin(v)*sin(u)/(2*pi) }
    <0,0>, <2*pi,2*pi>
    contained_by { box{-2, 2} }
    max_gradient 0.1
    accuracy 0.01
    precompute 10 x,y,z
      texture {pigment {color Red}
          finish {phong 1}
      }      }
```

Fig. 7.55 Rendering parametrically defined object

code defining it is given in Fig. 7.56. The code contains *File Header, Transform Node, Shape Node, Appearance Node, Geometry Node, Light source, and Background Nodes*. Every VRML file starts with the header *#VRML V2.0 utf8*. The *utf8* specification refers to an ISO standard for text strings known as UTF-8 encoding. Geometry of the shape is defined in the *geometry* node. The *Transform* node is a grouping node that defines *Translation, Rotation,* and *Scaling* transformations. In the example, it is the rotation about axis *X* by 0.7 radians. The *Shape* node is the basic container node for a geometry object. The *Appearance* node defines color, the smoothness, and the shininess of the surface, etc. This is done with the *Material* and texturing nodes. Colors are defined as red, green, and blue components in the domain [0, 1]. In the example, purple and shiny material is defined. Finally, a light source and a background color are defined.

Complex transformations can be constructed with *Translation, Rotation,* and *Scaling* transformations. The basic shapes available in VRML are *Sphere, Box, Cylinder, Cone, Face Set, Elevation Grid, Extrusion, NURBS curve* and *NURBS Surface, Polyline,* and *Text*. VRML also has tools for hierarchical representation of shapes and their rapid visualization by using references to the objects, level-of-detail, and visibility limit. Each object in VRML can be "anchored" to a URL pointing to a web page or another VRML object.

Other interactive tools are proximity and touch sensors which can be associated with objects, and an ability to create and process events differently, including using Java scripts.

The VRML code can be created and edited with any text editor, and it has to be saved with extension *wrl*. To visualize it, you need one of the VRML viewers to be installed on your computer. These viewers are often installed as plug-ins to internet browsers. The VRML scene defined in Fig. 7.56 is visualized by *BS Contact VRML* [10]. It can be downloaded from the book's software repository [6] *Software/FVRML/BSContactViewer.exe*. After the installation, the file with the extension *wrl* will be visualized when it is clicked at. The VRML files can be edited with any text editor or dedicated editors like *VrmlPad* [11].

```
#VRML V2.0 utf8
# Transform node may define 3 transformations: scaling, rotation
# and translation. They will be applied in exactly this order
# regardless the order in which they are written in the node
# Transform nodes can be nested to add more transformations and
# to change their order.
Transform {
      rotation  1 0 0 0.7
      translation    0 0 0
      scale 1 1 1
      children[


# Shape node includes geometry and appearance nodes.
      Shape {
# Purple color with 50% of shininess is defined
          appearance Appearance {
          material Material    {
              diffuseColor   0.5 0 0.5
              shininess      0.5 } }
# Geometry is defined as a cylinder with radius 3 and height 6.
# The centre of the cylinder is at the origin.
# Top, bottom and side polygons can be removed if FALSE is
defined.
          Geometry Cylinder  {
              radius 3  height 6
              side TRUE  top TRUE  bottom TRUE
          }
              }
      ]
      }
# Point light source with
# white color is defined
# at point with coordinates
# 250 400 150. It can be
# visible within 1500 m
# distance.
PointLight {
      on    TRUE
      ambientIntensity    1
      color      1 1 1
    location 250 400 150
      radius 1500    }
# Blue background color is
# set for the scene.
Background { skyColor 0 0 1 }
```

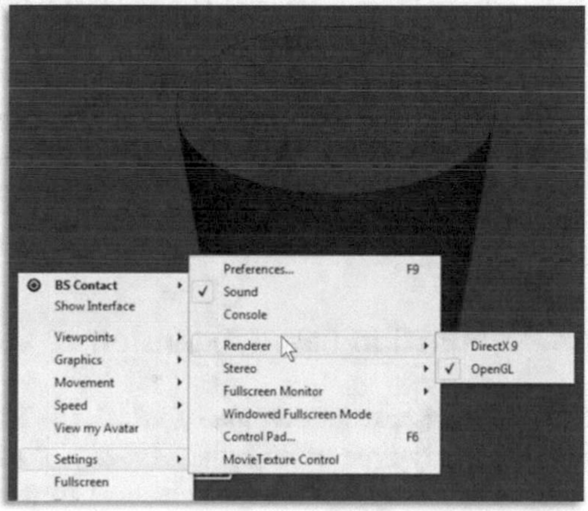

Fig. 7.56 A VRML code defining a cylinder and its visualization with BS Contact VRML/X3D viewer

7.4.2 Introduction to X3D

The Extensible 3D (X3D) [12] is a successor to VRML for real-time communication across various applications and networks. It includes features used in engineering and scientific visualization, CAD, multimedia and entertainment, etc. X3D has been already widely accepted as the successor to VRML. It introduces many new features to enhance the VRML meanwhile still provides backward compatibility with VRML files. It is suitable for various purposes, such as immersive virtual environment using special devices and visualization on clusters. A free authoring tool X3D-edit is also available to build and visualize the structure of X3D scenes [13].

For modeling and visualization purposes, X3D offers many types of geometries, appearances, and rigid transformations (i.e. rotation, translation, and scaling). The geometries include 2D primitives (such as points, arcs, circles, ellipses, lines, and polygons) and 3D primitives (such as boxes, cones, cylinders, and spheres). If complex 3D geometries are desired, polygon meshes can be employed. Besides primitives for general modeling purposes, specialized components, such as Humanoid Animation and NURBS for CAD applications, exist in X3D as well. The appearances in X3D include traditional 2D image textures, per-vertex colors, as well as many new techniques, such as multi-texturing, programmable shaders, and 3D color textures.

X3D files have extensions *x3d*, and the X3D scenes can be visualized by the same way as VRML scenes. The same *BS Contact Viewer* can be used for their visualization.

VRML and X3D are very versatile programming tools but they only visualize a few predefined shapes (sphere, cylinder, cone, box) and polygon meshes—the shapes defined by polygons. Definitions of the polygons' vertices by their coordinates is a tedious task. Usually, third-party software tools are used to generate these coordinates.

7.4.3 Function-Based Extension of VRML

The *Function-based Extension of VRML* (FVRML) allows for including in VRML practically any type of object's geometry, sophisticated graphics appearance, and transformation by writing mathematical formulae directly in the VRML code. It adds 2 new nodes—*FShape* and *FTransform*—to the standard VRML (Fig. 7.57).

The FShape node contains other 4 nodes in the same way how the standard VRML Shape node is organized. These nodes are FGeometry and FAppearance, which also contains FMaterial and FTexture3D nodes. These nodes can be used alone as well as together with the standard VRML nodes.

The FShape node is a container for the FGeometry or standard VRML *geometry* indexed nodes (IndexedFaseSet, IndexedLineSet), and the FAppearance or the standard Appearance node. These nodes define the geometry and the appearance of the shape, respectively.

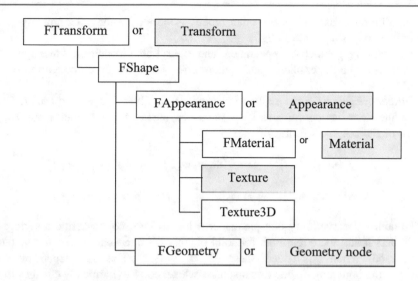

Fig. 7.57 Using extended and standard nodes of VRML. The standard nodes are shaded

The FGeometry node is designed to define the geometry using functions typed straight in the code as individual formulas and java-style function scripts or stored as algorithmic procedures in DLL libraries. If the standard *geometry* node is used in place of FGeometry node (only indexed geometry nodes can be used), the standard shapes of VRML will be assigned an appearance defined in FAppearance node.

The FAppearance node may contain the FMaterial or the standard VRML Material nodes, as well as the standard color Texture and the function-based FTexture3D nodes. In the FMaterial node, the components of the illumination model are defined using functions in a way how it can be done in the FGeometry node, i.e. in mathematical formulas, scripts, or DLLs. FMaterial node defines diffuse, specular, and emissive colors, as well as transparency, ambient intensity, and shininess by parametric and explicit functions. Parametric functions directly define color values *r*, *g*, *b*. When the color is defined by an explicit function, its values are linearly mapped to the *r*, *g*, *b* values of the actual color with a user-defined color interpolation map.

The FTexture3D node contains a displacement function for the geometry defined in the FGeometry node. The displacement function can be defined either as an explicit function of three Cartesian coordinates or as three parametric functions.

The FShape node may be called from the FTransform node or from the standard Transform node. The FTranform node defines affine transformation, which can be functions of time, as well as any other operations over the function-defined shapes. These are predefined set-theoretic (Boolean) intersection, union, difference, as well as any other function-defined set-theoretic and any other operations, e.g. shape morphing.

For defining geometry, appearance, and their transformations, three types of functions—implicit, explicit, and parametric—can be used concurrently in FVRML.

Parametric functions are explicit functions of variables u, v, w, and time t. They can define Cartesian coordinates x, y, z of curves, surfaces, solid objects, and r, g, b values of the colors as follows:

$$x = f_1(u|, v|, w|, t); \quad y = f_2(u|, v|, w|, t); \quad z = f_3(u|, v|, w|, t)$$

$$r = \varphi_1(u|, v|, w|, t); \quad g = \varphi_2(u|, v|, w|, t); \quad b = \varphi_3(u|, v|, w|, t)$$

To define a curve, only one parameter u has to be used, to define a surface–2 parameters u and v are required, for solid objects–all three parameters u, v, w have to be used. When t is added, these objects will become time-dependent. For example, the bouncing up and down sphere, whose color dynamically changes from green to red, can be then defined as follows:

$$x = R \cos v \cos u$$
$$y = R \sin u + a \sin t\pi$$
$$z = R \sin v \cos u$$
$$r = \sin t\pi$$
$$g = 1 - \sin t\pi$$
$$b = 0$$
$$u = [0, 2\pi], \quad v = [0, \pi], \quad t = [0, 1]$$

Implicit and explicit FRep functions are the functions defined as $f(x, y, z, t) = 0$ and $g = f(x, y, z, t) \geq 0$, respectively, where x, y, z are Cartesian coordinates and t is the time. They can only be used for defining 3D surfaces and solid objects since VRML is a 3D visualization system. Hence, a sphere can be defined by equation: $R^2 - x^2 - y^2 - z^2 = 0$. In the FVRML code, only the left part of the equality has to be written. As it was discussed in Chap. 2, by changing this equality into an *FRep* inequality $g = f(x, y, z, t) \geq 0$, we define not only a surface, but the space bounded by this surface, or a half-space. In this case, the function equals zero for the points located on the surface of the object, positive values of the function indicate points inside the solid object, and negative values are for the points which are outside the object. Let us consider an example of function $g = \sqrt{x^2 + y^2 + z^2}$ which defines a distance from the origin to any point with Cartesian coordinates (x, y, z). If we use function $g = R - \sqrt{x^2 + y^2 + z^2} \geq 0$, it will define a solid origin-centered sphere with radius R. The equation of a solid sphere could be also written as $g = R^2 - x^2 - y^2 - z^2 \geq 0$. Addition of time t to the parameters of the function will allow for making variable time-dependent shapes. For example, a sphere bouncing up and down by height a during time $t = [0,1]$ can be defined as $g = R^2 - x^2 - (y-a\sin(t\pi))^2 - z^2 \geq 0$. Implicit functions can efficiently represent Set-theoretic (Boolean) operations. In FVRML code, only the left part of the inequality (FRep) has to be written.

Geometry and color can be defined by implicit, explicit, or parametric functions in their own domains and then merged together into one shape. For example, we can define an origin-centered sphere with radius of 0.7 by the following implicit function:

$$0.7^2 - x^2 - y^2 - z^2 = 0$$

Then, a parametrically defined color is applied to it:

$$r = 1 \quad g = \text{abs}(\sin(u)) \quad b = 0$$

The final shape created with these implicit and parametric functions and its FVRML code is shown in Fig. 7.58. The first part of this code, beginning with EXTERPROTO, defines a prototype of the function-based VRML plug-in—declarations of the variables used. Only a little part of this code is shown in the Figure to save space. The EXTERNPROTO code is the same for any FVRML code. Note also that only the left part of the implicit function has to be typed in the code since it is always assumed to be defined as greater than or equal to 0.

Defining complex shapes usually assumes using multiple formulas and temporary variables. This requires a script-like mathematical language. FVRML emulates a subset of JavaScipt in which all variables, arrays, and constants have only one type *float*. The following mathematical functions are implemented:

$$\text{abs}(x), \; \text{fabs}(x), \; \text{sqrt}(x), \; \text{exp}(x), \; \log(x), \; \sin(x), \; \cos(x), \; \tan(x), \; \text{acos}(x), \; \text{asin}(x),$$
$$\text{atan}(x), \; \text{ceil}(x), \; \text{floor}(x), \; \text{atan2}(y,x), \; \text{mod}(x,y), \; \text{round}(x), \; \max(x,y),$$
$$\min(x,y), \; \cosh(x), \; \sinh(x), \; \tanh(x), \; \log10(x).$$

There are also flow control operators: *for* − *loops*, *while* − *loops*, *do* − *while* − *loops*, *break*, *continue*, and *if* − *else*. When writing function scripts, implicit functions $f(x,y,z) = 0$ and explicit functions $f(x,y,z) \geq 0$ have to be named *'function frep'*. Parametric functions for shapes have to be named *'function parametric_x'*, *'function parametric_y'*, and *'function parametric_z'*. Variables, x, y, z are reserved for Cartesian coordinates, while variables u, v, w are parameters. Variable t is reserved for defining the time.

Self-explanatory examples of defining parametric curve, surfaces, solid objects, implicit surfaces, constructive solid geometry, and animated shape morphing are given in Figs. 7.59, 7.60, 7.61, 7.62, 7.63, 7.64 and 7.65.

FVRML files have to be defined with *wrl* extension, as the common VRML files. They are edited and visualized in the same way as VRML files. Before that, the extension plug-in *Software/FVRML/FNode.exe* from the book's software repository has to be installed.

```
#VRML V2.0 utf8
EXTERNPROTO FGeometry [
    exposedField SFString definition
    exposedField MFFloat parameters
    ............................. . .
[ the rest of the EXTERNPROTO is skipped]

FShape {
    geometry FGeometry {
        definition "0.7^2-x^2-y^2-z^2"
        bboxCenter 0 0 0
        bboxSize 1.4 1.4 1.4
        resolution [75 75 75]
}

    appearance FAppearance { material FMaterial {
        diffuseColor "r=1; g=abs(sin(u*pi)); b=0;"   }      }
}
```

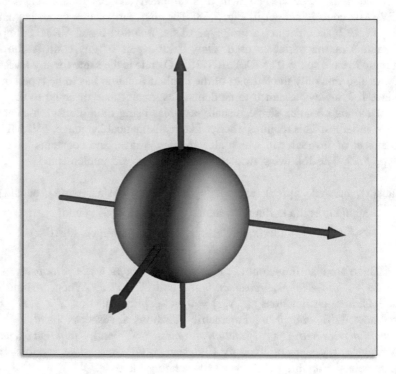

Fig. 7.58 A VRML code defining a cylinder and its visualization with BS Contact

```
#VRML V2.0 utf8
EXTERNPROTO FGeometry [
    exposedField SFString definition
    exposedField MFFloat parameters
    ...................................... . .
[ the rest of the EXTERNPROTO is skipped]

# External VRML object "Coordinate Axes" is included in the scene.
# The size of the axes can be changed by the scale transformation
Transform {
        scale 1.2 1.2 1.2 children [
            Inline {url "CoordinateAxes.wrl"}  ]}

FShape {
# This definition is needed for drawing curves
polygonizer      "analytical_curve"

geometry FGeometry {
# The parametric formulas defining the curve.
   definition "
   x = 0.75*sin(4*u*PI - 0.5*PI)*cos(2*u*PI) + 1.0;
   y = 0.75*sin(4*u*PI - 0.5*PI)*sin(2*u*PI);
   z = 0;"

# Domain for the parameter u.
parameters [0 1]
# Sampling resolution
# along the curve. It
# defines how many times
# the parameter domain
# is sampled to
# calculate the curve
# function.
resolution [100]
  }

appearance FAppearance {
material FMaterial {
# Fixed red color is
# defined for the curve.
diffuseColor
"r=1; g=0; b=0;"
    }     }
}
```

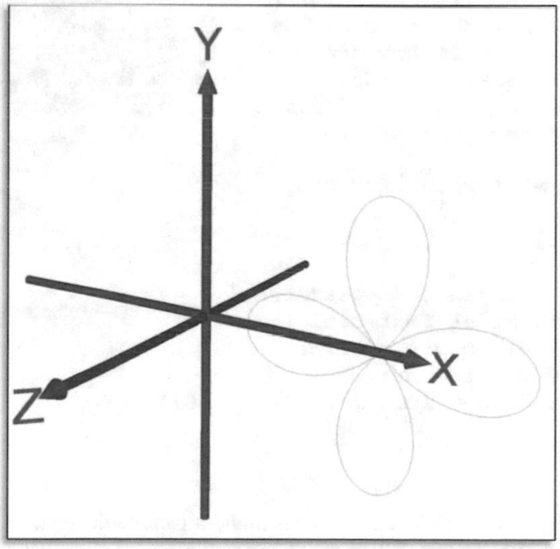

Fig. 7.59 An FVRML code defining a parametric curve which was previously displayed using OpenGL in Fig. 7.5

```
#VRML V2.0 utf8
EXTERNPROTO FGeometry [
    exposedField SFString definition
    exposedField MFFloat parameters
    .............................. . .
[ the rest of the EXTERNPROTO is skipped]

# External VRML object "Coordinate Axes" is included in the scene.
# The size of the axes can be changed by the scale transformation
Transform {
      scale 1.2 1.2 1.2 children [
          Inline {url "CoordinateAxes.wrl"}  ]}

FShape {
geometry FGeometry {

# The parametric formulae defining the surface.
definition "
x=(0.75*sin(4*u*PI-0.5*PI)*cos(2*u*PI)+1.0)*sin(v*3*PI/2+PI/2);
y=0.75*sin(4*u*PI-0.5*PI)*sin(2*u*PI)+1.0*v;
z=(0.75*sin(4*u*PI-0.5*PI)*cos(2*u*PI)+1.0)*cos(v*3*PI/2+PI/2);

# The parameters domain.
parameters [0 1 0 1]

# Sampling resolution in
# parameters u and v.
# This is how the
# parameters domain is
# sampled to calculate
# the geometry function.

resolution [200 200]

 }

appearance FAppearance {
material FMaterial {
# Fixed green color is defined
# for the surface.
diffuseColor "r=0; g=1; b=0;"
    }     }
}
```

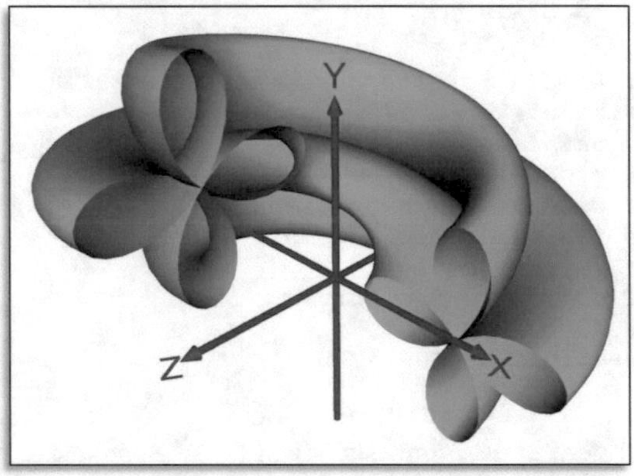

Fig. 7.60 An FVRML code defining a parametric surface which was previously displayed using OpenGL in Fig. 7.7

7.4.4   Function-Based Extension of X3D

In terms of shape modeling features, X3D has many benefits, and is widely used in many industrial areas. However, it still has several deficiencies. Though it provides advanced polygon-based representation methods such as triangle fans and strips, as well as compressed binary encoding, the file sizes would still be large for complex

```
#VRML V2.0 utf8
EXTERNPROTO FGeometry [
    exposedField SFString definition
    exposedField MFFloat parameters
    ................................. . .
[ the rest of the EXTERNPROTO is skipped]

# External VRML object "Coordinate Axes" is included in the scene.
# The size of the axes can be changed by the scale transformation
Transform {
        scale 1.2 1.2 1.2 children [
            Inline {url "CoordinateAxes.wrl"}  ]}


FShape {
geometry FGeometry {

# The parametric formulae defining the surface.
definition "x=(0.3*sin(10*(u*pi+v*pi))+1)*cos(u*pi);
            y=(0.3*sin(10*(u*pi+v*pi))+1)*sin(u*pi)*cos(v*pi);
            z=(0.3*sin(10*(u*pi+v*pi))+1)*sin(u*pi)*sin(v*pi);"

# The parameters domain.
parameters [-1 1 -1 1]

# Sampling resolution in
# parameters u and v.
# This is how the parameters
# domain is sampled to
# calculate the
# geometry function.

resolution [200 200]

 }
appearance FAppearance {
material FMaterial {
# rgb values of color are
# computed by mapping a
# distance function
# values 0.7 and 1.3 to the
# colors 0 1 0 and 1 0 0  and
# interpolating color between
# them
diffuseColor "sqrt(x*x+y*y+z*z)"
patternKey [0.7 1.3]
patternColor [0 1 0   1 0 0]
    }     }
}
```

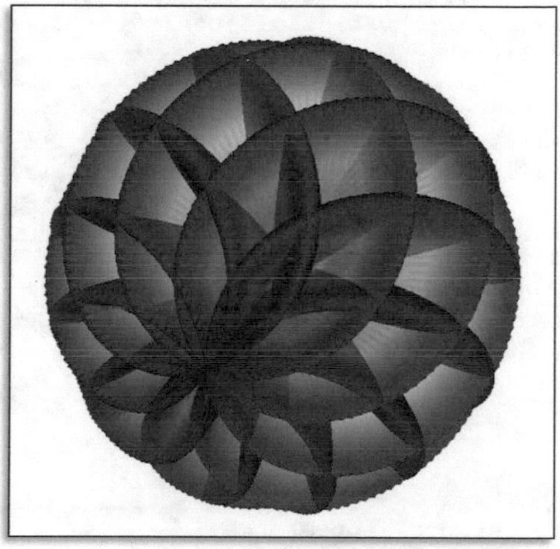

Fig. 7.61 An FVRML code defining a parametric surface

shapes. Also, without commercial modeling software, Constructive Solid Geometry (CSG) is not supported by X3D. Even with commercial software, CSG can only be done at the level of polygon meshes. Finally, it is not easy for authors to apply 3D

```
#VRML V2.0 utf8
EXTERNPROTO FGeometry [
    exposedField SFString definition
    exposedField MFFloat parameters
    ............................ . .
[ the rest of the EXTERNPROTO is skipped]

# External VRML object "Coordinate Axes" is included in the scene.
# The size of the axes can be changed by the scale transformation
Transform {
        scale 1.2 1.2 1.2 children [
        Inline {url "CoordinateAxes.wrl"}  ]}

FShape {
geometry FGeometry {

# The parametric formulae defining the solid.
definition "
x=((0.35+w*0.5)*sin(4*u*PI-0.5*PI)*cos(2*u*PI)+1.0)*sin(v*3*PI/2
+PI/2);
y=(0.35+w*0.5)*sin(4*u*PI-0.5*PI)*sin(2*u*PI)+v;
z=((0.35+w*0.5)*sin(4*u*PI-0.5*PI)*cos(2*u*PI)+1.0)*cos(v*3*PI/2
+PI/2);"
```

```
# The parameters domain.
parameters
[0 1 0 1 0 1]

# Sampling resolution
# in parameters
# u, v and w.
resolution
[100 100 100]
  }

appearance FAppearance {
material FMaterial {
# Fixed green color is
# defined for the surfaces
# bounding the solid object
diffuseColor "r=0; g=1; b=0;"
    }      }
      }
```

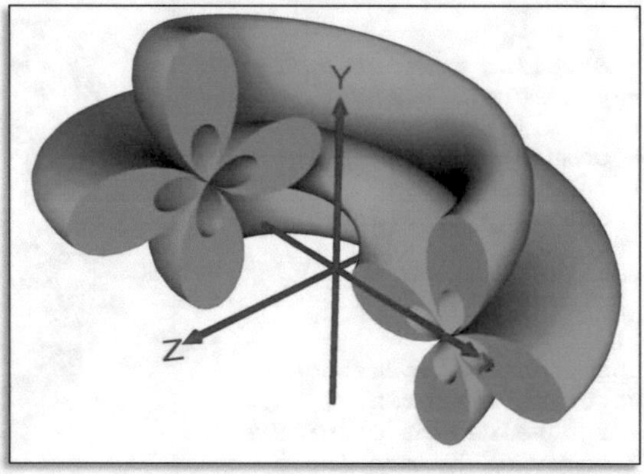

Fig. 7.62 An FVRML code defining a parametric solid object which was previously displayed using OpenGL in Fig. 7.13

```
#VRML V2.0 utf8
EXTERNPROTO FGeometry [
    exposedField SFString definition
    exposedField MFFloat parameters
    .............................. . .
[ the rest of the EXTERNPROTO is skipped]

Background {skyColor 1 1 1}

# External VRML object "Coordinate Axes" is included in the scene.
# The size of the axes can be changed by the scale transformation
Transform {
     scale 1.2 1.2 1.2 children [
          Inline {url "CoordinateAxes.wrl"}  ]}


FShape {
geometry FGeometry {

# Function script defining the surface.
    definition "x+y+z"

# Adjust the tight bounding box and an optimal resolution
bboxCenter 0 0 0
bboxSize 2 2 2

# Sampling resolution in
# three dimensions
resolution [100 100 100]

  }

appearance FAppearance {
material FMaterial {
# Variable linear change
# of green color from
# 0 to 1 is defined for
# the surface within the
# bounding box
# x, y, z = [-1 1]
diffuseColor
"r=1; g=(v+1)/2; b=0;"
    }    }
}
```

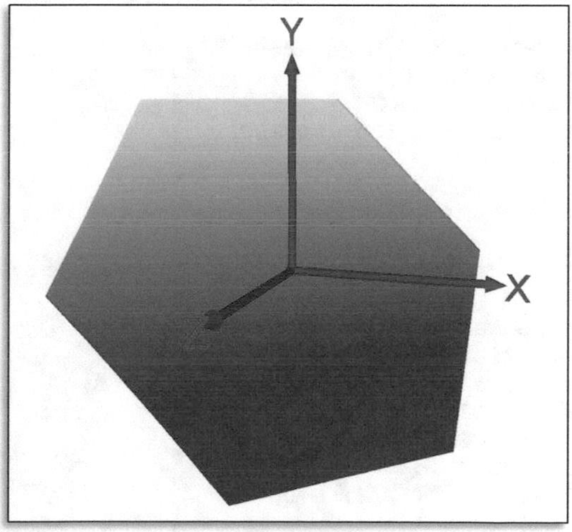

Fig. 7.63 An FVRML code defining an implicit surface

geometric textures to a 3D geometry, and the animation support in X3D is still limited, despite some improvements which have been done. These problems can be solved by using FX3D—the function-based extension of X3D.

Figure 7.66 is an example of FX3D modeling which illustrates the ability of a concurrent definition of the time-dependent geometry, colors, and geometric textures. In this example, geometry, colors, and geometric textures are all defined by

```
#VRML V2.0 utf8
EXTERNPROTO FGeometry [
    exposedField SFString definition
    exposedField MFFloat parameters
..............................................
[ the rest of the EXTERNPROTO is skipped]

# External VRML object "Coordinate Axes" is included in the scene.
# The size of the axes can be changed by the scale transformation
Transform {
      scale 1.2 1.2 1.2 children [
      Inline {url "CoordinateAxes.wrl"}  ]}

FShape {
geometry FGeometry {

# Function script defining the CSG solid. Change to some other
# formulae to see how the solid geometry changes
# based on the parameters domain and the sampling resolution
# defined below
    definition "      function frep(x,y,z,t){
                superellipsoid=0.7^6-x^6-y^6-z^6;
                cylinder=0.25^2-x^2-y^2;
                final=min(superellipsoid, -cylinder);
                return final;}"

# Adjust the tight bounding box and an optimal resolution
bboxCenter 0 0 0
bboxSize 2 2 2
resolution [100 100 100]
  }
```

```
appearance FAppearance {
material FMaterial {
# Variable linear change
# of color is defined for
# the CGS solid
diffuseColor
"r=1;  g=(v+1)/2; b=0;"
      }      }
}
```

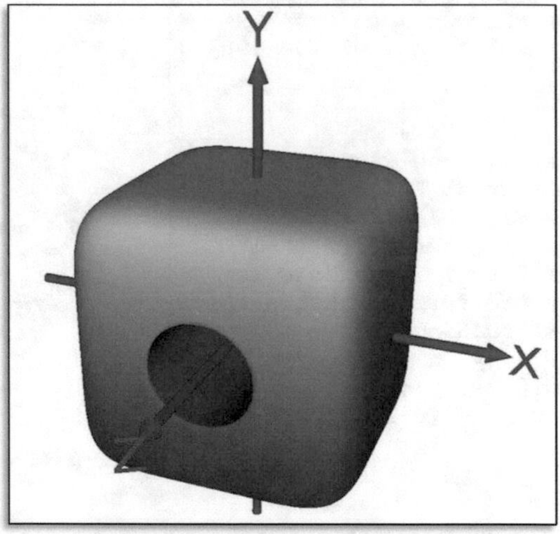

Fig. 7.64 An FVRML code defining an FRep solid object

the time-dependent functions with different time spans. The animated scene shows rippling waves which are distorted by solid noise. The color of the waves changes through time.

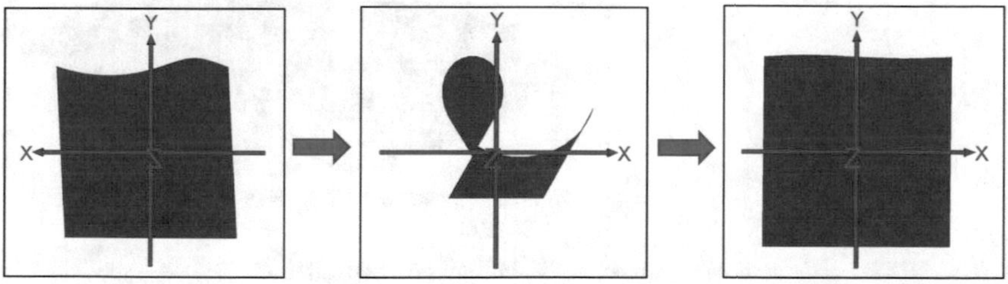

```
#VRML V2.0 utf8
EXTERNPROTO FGeometry [
    exposedField SFString definition
    exposedField MFFloat parameters
    ...............................................  . .
```
[the rest of the EXTERNPROTO is skipped]
 # External VRML object "Coordinate Axes" is included in the scene.
 # The size of the axes can be changed by the scale transformation
Transform {
        scale 1.2 1.2 1.2 children [
            Inline {url "CoordinateAxes.wrl"}]}
FShape {
 # Enabling cycled animation
loop TRUE
 # Mapping the interval of the internal time t=[0,1] to the actual
time in sec.
cycleInterval 7
geometry FGeometry {
resolution [30 30]
parameters [0 1 0 1]
 # Definition of the animated linear transformation (morphing)
# of a square polygon defined by x1(u,v), y1(u,v), z1(u,v)
 # to a circular disk defined by x2(u,v), y2(u,v), z2(u,v)
definition "
function parametric_x(u,v,w,t)
 # linear morphing function for x-coordinate
    { x1=-1+2*u;   x2=v*cos(u*2*pi);   return x1+(x2-x1)*t; }
function parametric_y(u,v,w,t)
 # linear morphing function for y-coordinate
    { y1=-1+2*v;   y2=v*sin(u*2*pi);   return y1+(y2-y1)*t; }
function parametric_z(u,v,w,t)
 # linear morphing function for z-coordinate
{return 0;}"
 }
appearance **FAppearance** {
material **FMaterial** {
diffuseColor "r=1; b=0; g=0;"
}   }
}
```

**Fig. 7.65** An FVRML code defining parametric morphing

Figure 7.67 shows another FX3D example where an animated shape is created by sweeping a parametric curve along another parametric curve.
```

```
<!--Shape-->
<ProtoInstance name="FShape">
  <fieldValue name="cycleInterval" value="5" />
  <fieldValue name="loop" value="true" />
  <fieldValue name="appearance">

    <ProtoInstance name="FAppearance">
      <fieldValue name="material">

        <ProtoInstance name="FMaterial">
          <fieldValue name="diffuseColor" value=
"r=cos(v*pi*2+t*pi*6);g=fabs(sin(u));b=sin(v*pi*2+t*pi*4)/2;" />
        </ProtoInstance>
      </fieldValue>
      <fieldValue name="texture3D">

        <ProtoInstance name="FTexture3D">
          <fieldValue name="timeSpan" value="0 0.4" />
<fieldValue name="definition" value="fabs(t-0.2)*(sin(2*pi*x)
*sin(2*pi*y)+sin(2*pi*x)*sin(2*pi*z)+sin(2*pi*y)*sin(2*pi*z))"
/>
        </ProtoInstance>
      </fieldValue>
    </ProtoInstance>
  </fieldValue>
  <fieldValue name="geometry">
    <ProtoInstance name="FGeometry">
      <fieldValue name="resolution" value="50 100" />
  <fieldValue name="parameters" value="0 6.28 0 6.28 -1 1" />
      <fieldValue name="definition" value="
function parametric_x(u,v,w,t)
{return v*5*cos(u);}
function parametric_y(u,v,w,t)
{return v*5*sin(u);}
function parametric_z(u,v,w,t)
{return cos(v*5+t*2*pi);}"/>
    </ProtoInstance>
  </fieldValue>
</ProtoInstance>
```

Time-dependent color

Time-dependent geometric texture

Time-dependent geometric shape

Fig. 7.66 FX3D code combining shape, color, and texture animation

```
<ProtoInstance name="FShape">
 <fieldValue name="cycleInterval" value="5" />
 <fieldValue name="loop" value="true" />
 <fieldValue name="appearance">

  <ProtoInstance name="FAppearance">
   <fieldValue name="material">
    <ProtoInstance name="FMaterial">
     <fieldValue name="timeSpan" value="-3.1415926 3.1415926" />

     <fieldValue name="diffuseColor"
                  value="sin(sqrt(x*x+y*y+z*z) * pi + t)" />
     <fieldValue name="patternColor" value="0 1 0 1 0 0" />
     <fieldValue name="patternKey" value="-1 1" />

    </ProtoInstance>
   </fieldValue>
  </ProtoInstance>
 </fieldValue>
```

Time-dependent color

```
 <fieldValue name="geometry">
  <ProtoInstance name="FGeometry">
   <fieldValue name="resolution" value="200 30" />
   <fieldValue name="parameters" value="-1 1 -1 1 -1 1" />
   <fieldValue name="timeSpan" value="0.01 1" />

 <fieldValue name="definition" value="
  function parametric_x(u,v,w,t) {u=(u+1)*pi*2*t;v=(v+1)*pi;
  return 0.1*cos(v)*cos(u)+0.5*cos(u)*(1+0.4*cos(1.5*u));}
     function parametric_y(u,v,w,t) {u=(u+1)*pi*2*t;v=(v+1)*pi;
     return 0.1*sin(v)+0.4*sin(1.5*u)-1;}
  function parametric_z(u,v,w,t) {u=(u+1)*pi*2*t;v=(v+1)*pi;
  return 0.1*cos(v)*sin(u)+0.5*sin(u)*(1+0.4*cos(1.5*u))-1;}" />
  </ProtoInstance>
 </fieldValue>
</ProtoInstance>
```

Time-dependent geometric shape

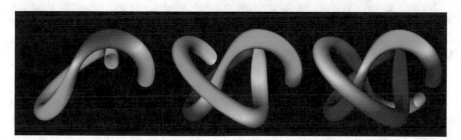

Fig. 7.67 Animated shape metamorphosis in FX3D

FX3D files have to be defined with *wrl* extension, as the VRML and FVRML files. They are edited and visualized in the same way as FVRML files. Before that, the extension plug-in *Software/FVRML/FNode.exe* has to be installed from the book's software repository.

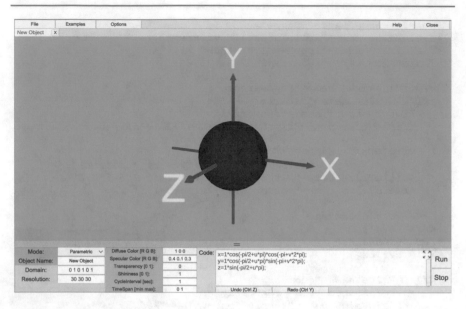

Fig. 7.68 Shape Explorer

7.5 Drawing with Shape Explorer

Shape Explorer (Fig. 7.68) is an interactive software tool designed to use the same function-based definitions for shapes: implicit and parametric formulas, domains and sampling resolutions. It is designed to run both on Windows and Mac. It can be downloaded from the book's software repository for Windows and Mac from *Software/ShapeExplorer*.

Shape Explorer is just one interactive window where you can type the function definitions and other parameters. The shape definitions can be saved in a proprietary format and loaded later to the software to continue working with it. Migration between Shape Explorer and FVRML can be done by copy-pasting the formulae, domains, resolutions, and colors. Shape Explorer supports only one shape visualization at the time. It does not provide any transformations and other advanced features available in VRML. Its purpose is to work as a quick all-in-one multi-platform visualization tool. Figures 7.69, 7.70, 7.71 and 7.72 show the making of the shapes which were previously displayed using OpenGL and FVRML.

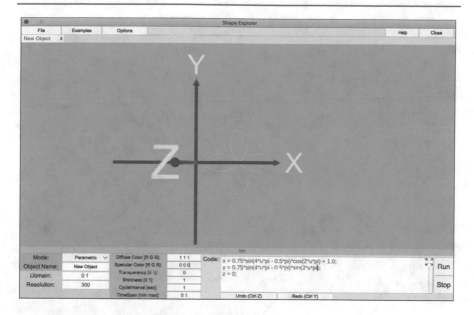

Fig. 7.69 Displaying a parametric curve as in Fig. 7.59

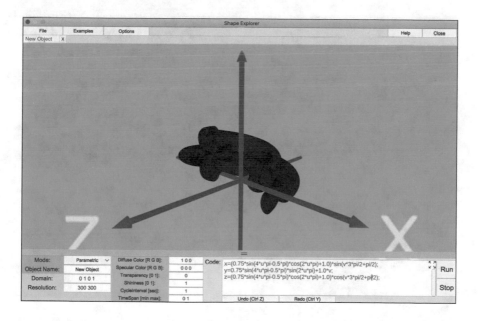

Fig. 7.70 Displaying a parametric surface as in Fig. 7.60

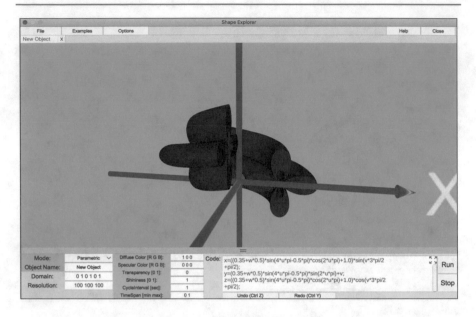

Fig. 7.71 Displaying a parametric solid as in Fig. 7.62

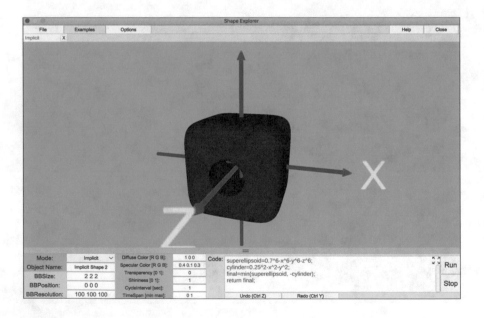

Fig. 7.72 Displaying an FRep solid as in Fig. 7.64

7.6 Summary

- The computer graphics and visualization software tools are given to us as software libraries which are extensions of common programming languages, as well as specially developed stand-alone graphics libraries and interactive systems.
- The graphics and visualization software tools can be classified by the programming style. The imperative style of programming tells the computer how to do things, while when we use the declarative style of programming, we tell the computer what to do.
- The same principles of separation of geometry and visual appearance are used in many computer graphics libraries and tools.
- Polygon-based visualization (e.g. by OpenGL) is fast but it compromises on precision of presentation. Geometry of fine details is often replaced by image textures (patterns) displayed on the surfaces.
- Ray tracing (e.g. by POV-Ray) is very precise but slow. It is mostly used for making images and not designed to be an interactive visualization tool.
- VRML/X3D is polygon-based but declarative style virtual scene description language. Compared to OpenGL and other graphics libraries, it requires very little time to start programming for both web-enabled and local visualization problems. It is full of technological solutions but may require a deeper learning to master them.
- FVRML/FX3D is an extension of VRML and X3D which allows to use mathematical functions and procedures for defining geometry and appearance f the objects.
- Shape Explorer is an interactive software tool designed to use function-based definitions for shapes: implicit and parametric formulas, domains and sampling resolutions. It is designed to run both on Windows and Mac.

References

1. OpenGL web page, https://www.khronos.org/opengl/
2. GLUT web page, https://www.xmission.com/ ~ nate/glut.html
3. FreeGLUT, https://freeglut.sourceforge.net
4. OpenGL manual, https://www.khronos.org/registry/OpenGL-Refpages//
5. GLUT manual, https://www.opengl.org/resources/libraries/glut/glut-3.spec.pdf
6. Book software repository, https://personal.ntu.edu.sg/assourin/BookSpringer or https://drive.google.com/drive/folders/1bSzOhXRpbxNzylStk00-HZA-QL3-5aK7?usp=sharing
7. POV-Ray, https://www.povray.org
8. Moray, https://www.stmuc.com/moray
9. VRML specification, https://www.web3d.org/x3d/vrml

10. BS Contact VRML, https://www.bitmanagement.de
11. VrmlPad, https://www.parallelgraphics.com/products/vrmlpad/download/
12. Extensible 3D (X3D), https://www.web3d.org/x3d/specifications
13. Don Brutzman and Leonard Daly (2007), *X3D: Extensible 3D Graphics for Web Authors*, Elsevier.

Index

Printed in the United States
by Baker & Taylor Publisher Services